Cosmology, Atomic Theory, Evolution

Classic Readings in the Literature of Science

Edited by
William C. Dampier
and
Margaret Dampier

DOVER PUBLICATIONS, INC.
MINEOLA, NEW YORK

Bibliographical Note

This Dover edition, first published in 2003, is an unabridged republication of the 1959 Harper Torchbook reprint of the work which was originally published in 1924 by the Cambridge University Press, Cambridge (England) under the title, *Cambridge Readings in the Literature of Science: Being extracts from the writings of men of science to illustrate the development of scientific thought.*

Library of Congress Cataloging-in-Publication Data

Dampier, William Cecil Dampier, Sir, 1867-1952.
Cosmology, atomic theory, evolution : classic readings in the literature of science / arranged by William C. Dampier and his daughter Margaret Dampier.
p. cm.
Originally published: Cambridge readings in the literature of science. Cambridge : Cambridge University Press, 1924.
Includes index.
ISBN 0-486-42805-2 (pbk.)
1. Science. I. Dampier, Margaret. II. Dampier, William Cecil Dampier, Sir, 1867-1952. Cambridge readings in the literature of science. III. Title.

Q171.D28 2003
509--dc21

2003043461

Manufactured in the United States of America
Dover Publications, Inc., 31 East 2nd Street, Mineola, N.Y. 11501

PREFACE

In arranging a book of extracts from the writings of men of science, two methods are possible. Passages may be chosen for their intrinsic interest or for their importance in the history of knowledge, but isolated from the less outstanding work that led up to and followed them. On the other hand, extracts may be taken to illustrate the development of definite subjects in the thought of succeeding ages.

This volume is planned on the second method to tell a connected story. We have picked out as threads on which to string our anthology of science the ideas of mankind on three problems of transcending importance:

(1) the structure of the universe—cosmogony;
(2) the nature of matter—atomic theories;
(3) the development of life—evolution.

Along these three lines we try to trace the thoughts of man from the inspired poetry of the Book of Genesis to the latest revelations of the telescope and the laboratory.

W.C.D.
M.D.

CONTENTS

I. COSMOGONY

II. THE ATOMIC THEORY

III. THE THEORY OF EVOLUTION

CONTENTS

Cosmology, Atomic Theory, Evolution

I. COSMOGONY

THE BOOK OF GENESIS

CHAPTER I, *and* CHAPTER II, *verses* 1–3

IN the beginning God created the heaven and the earth.

And the earth was without form and void; and darkness was upon the face of the deep. And the Spirit of God moved upon the face of the waters.

And God said, Let there be light: and there was light.

And God saw the light, that it was good: and God divided the light from the darkness.

And God called the light Day, and the darkness he called Night. And the evening and the morning were the first day.

And God said, Let there be a firmament in the midst of the waters, and let it divide the waters from the waters.

And God made the firmament, and divided the waters which were under the firmament from the waters which were above the firmament: and it was so.

And God called the firmament Heaven. And the evening and the morning were the second day.

And God said, Let the waters under the heaven be gathered together unto one place, and let the dry land appear: and it was so.

And God called the dry land Earth; and the gathering together of the waters called he Seas: and God saw that it was good.

And God said, Let the earth bring forth grass, the herb yielding seed, and the fruit tree yielding fruit after his kind, whose seed is in itself, upon the earth: and it was so.

And the earth brought forth grass, and herb yielding seed after his kind, and the tree yielding fruit, whose seed was in itself, after his kind: and God saw that it was good.

And the evening and the morning were the third day.

And God said, Let there be lights in the firmament of the heaven to divide the day from the night; and let them be for signs, and for seasons, and for days, and years:

And let them be for lights in the firmament of the heaven to give light upon the earth: and it was so.

And God made two great lights; the greater light to rule the day, and the lesser light to rule the night: he made the stars also.

And God set them in the firmament of heaven to give light upon the earth,

And to rule over the day and over the night, and to divide the light from the darkness: and God saw that it was good.

And the evening and the morning were the fourth day.

And God said, Let the waters bring forth abundantly the moving creature that hath life, and fowl that may fly above the earth in the open firmament of heaven.

And God created great whales, and every living creature that moveth, which the waters brought forth abundantly, after their kind, and every winged fowl after his kind: and God saw that it was good.

And God blessed them, saying, Be fruitful, and multiply, and fill the waters in the seas, and let the fowl multiply in the earth.

And the evening and the morning were the fifth day.

And God said, Let the earth bring forth the living creature after his kind, cattle, and creeping thing, and beast of the earth after his kind: and it was so.

And God made the beast of the earth after his kind, and cattle after their kind, and every thing that creepeth upon the earth after his kind: and God saw that it was good.

And God said, Let us make man in our image, after our likeness: and let them have dominion over the fish of the sea, and over the fowl of the air, and over the cattle, and over all the earth, and over every creeping thing that creepeth upon the earth.

So God created man in his own image, in the image of God created he him; male and female created he them.

And God blessed them, and God said unto them, Be fruitful and multiply, and replenish the earth, and subdue it; and have dominion over the fish of the sea, and over the fowl of the air, and over every living thing that moveth upon the earth.

And God said, Behold, I have given you every herb bearing seed, which is upon the face of all the earth, and every tree, in the which is the fruit of a tree yielding seed; to you it shall be for meat.

And to every beast of the earth, and to every fowl of the air, and to every thing that creepeth upon the earth, wherein there is life, I have given every green herb for meat: and it was so.

And God saw everything that he had made, and, behold, it was very good. And the evening and the morning were the sixth day.

Thus the heavens and the earth were finished, and all the host of them.

And on the seventh day God ended his work which he had made; and he rested on the seventh day from all his work which he had made.

And God blessed the seventh day, and sanctified it; because that in it he had rested from all his work which God created and made.

ARISTOTLE

The beginnings of science can be traced in Babylonian astronomy and in Egyptian geometry and medicine. In Greece, the genius of a gifted race used the knowledge of Babylon and Egypt as a subject for more abstract thought. The writings of the earlier philosophers are seldom represented by more than isolated fragments, and our knowledge of their work is chiefly derived from references and quotations in later authors. Of these, the most important in the history of thought was Aristotle, who lived from about 384 to about 321 B.C. He had a share in the education of Alexander the Great, who afterwards supplied money to forward Aristotle's researches. Most of his works survive, and contain an encyclopædic study of the knowledge of his time. Perhaps Aristotle's greatest strength lay in biology, and there we shall meet him again. In astronomy and physics he was less successful. He attempted too much. The true line of immediate advance lay in the more limited but more exact methods of Aristarchus and Archimedes. Nevertheless, both for his own ideas and for an account of those of other Greek philosophers, Aristotle's physical works are of great interest.

Moreover, commentaries on Aristotle were almost the only channel by which the ancient learning passed through the dark ages in Western Europe, and the rediscovery of his works themselves marked the culmination of mediæval thought. It was only at the Renaissance that men began to see that discovery might pass beyond the knowledge of Aristotle, and only with the rise of modern experimental methods that his physics became obsolete; indeed the weight of his authority delayed for a time the acceptance of the new knowledge.

ON THE HEAVENS

(Some freely rendered extracts, based on the literal translation of Thomas Taylor, 1807.)

All men believe that there are gods, and all men, both barbarians and Greeks, assign the highest place in heaven to the divine nature....For, according to tradition, in the whole of past time no change has taken place either in the heaven as a whole or in any of its parts. Moreover, the name by which we call it appears to have been handed down in succession from the ancients, who held the same opinion about its divine nature which lasts to the present time.

For such reasons, then, we believe that the heaven was neither created nor is it corruptible, but that it is one and everlasting, unchanged through infinite time. Hence we may well persuade ourselves that ancient assertions, especially those of our own ancestors, are true, and see that one kind of motion is immortal and divine, having no end, but being itself the end of other motions. Now motion in a circle is perfect, having neither beginning nor end, nor ceasing in infinite time.

As the ancients attributed heaven and the space above it to the gods, so our reasoning shows that it is incorruptible and uncreated and untouched by mortal troubles. No force is needed to keep the heaven moving, or to prevent it moving in another manner;...nor need we suppose that its stability depends on its support by a certain giant Atlas, as in the ancient fable: as though forsooth all bodies on high possessed gravity and an earthly nature. Not thus has it been preserved for so long, nor yet, as Empedocles asserts, by whirling round faster than its natural motion downwards. Nor is it reasonable to think that it remains unchanged by the compulsion of a soul, untiring and sleepless, unlike the soul of mortal animals, for it would need the fate of some Ixion (bound for ever to a fiery wheel) to keep it in motion....

The heaven, moreover, must be a sphere, for this is the only form worthy of its essence, as it holds the first place in nature.... Every plane figure is contained by straight lines or by a circumference. The right-lined figure is bounded by many lines, but the circle by but one. But as the one is prior to the many and the simple to the composite, so the circle is the first of plane figures.... Again, to a straight line an addition can always be made, but to a circular line never. Thus once more the line which traces a circle is perfect. Hence if the perfect is prior to the imperfect, the circle again will be the first of figures. In like manner also the sphere will be the first of solids; for this alone is contained by one superficies, while flat-sided figures are contained by many. As a circle is in planes, so is a sphere in solids....

Further still, since it seems clear and we assume that the universe revolves in a circle, and since beyond the uttermost sky is neither body nor space nor vacuum, once more it follows that the universe is spherical. For, if it were rectilinear, there must be space beyond it: a rectilinear body as it revolves will never occupy the same place: where it formerly was it is not now, and where it is not now it will be again, because the corners project....

It remains to discuss the earth—where it is situated, whether it is at rest or moves, and what is its form. With regard to its position, all philosophers have not the same opinion. Most of those who assert that the heaven is finite say that the earth lies at the centre, while those in Italy who are called Pythagoreans hold the contrary. For they say that at the centre of the universe is fire, and that the earth being one of the stars moves in a circle about that centre and thus causes day and night. They also invent opposite to our earth another earth, which they call counter-earth: not investigating theories and causes to explain the facts, but adjusting the facts to fit certain opinions and theories of their own. To many others also it seems that a central place should not be assigned to the earth for reasons not based on facts but on opinions. For they fancy that the most honourable place belongs to the most honourable nature: that fire is more honourable than earth and the boundaries of a space than the region within. But the circumference and the centre they say are boundaries. So that, thus reasoning, they think not the earth but fire holds place in the centre of the sphere. Further still, the Pythagoreans hold that the chief place should be best guarded, and call the centre the altar of Zeus, and thus again assign this place to fire, as if the centre of a mathematical figure and the middle of a thing or the natural centre were of the same kind....Such as assert that the earth is not situated in the middle of the universe are of opinion that it and the counter-earth also move round the centre in a circle. And to some it appears that many such bodies may move round the centre though invisible to us by the intervention of the earth. Hence they say there are more eclipses of the moon than of the sun, for each of the moving bodies, and not the earth only, can obstruct the light of the moon....But some say that the earth, being situated in the centre, rolls round the pole which is extended through the universe, as it is written in the *Timaeus*.

In a similar way there is doubt about the shape of the earth. To some it seems to be spherical, but to others flat, in the form of a drum. To support this opinion they urge that, when the sun rises and sets, he appears to make a straight and not a circular occultation, as it should be if the earth were spherical. These men do not realise the distance of the sun from the earth and the magnitude of the circumference, nor do they consider that, when seen cutting a small circle, a part of the large circle appears at a distance as a straight line. Because of this appearance, therefore, they ought not to deny that the earth is round....It is indeed irrational not to wonder how it is that a small fragment of earth if dropt from a high place moves downward and a larger fragment more swiftly downward, while the whole earth does not tend downward and its great bulk is at rest. For if, while fragments of earth are falling, some one could take away the whole earth before they reached it, they would nevertheless move downward if nothing opposed them. Hence this question is of general philosophic interest, its consequences seeming no less difficult than the problem. For some on this account hold that the part of the earth below us must be infinite, as Xenophanes of Colophon says, rooted to infinity....Hence the rebuke of Empedocles when he writes:

The boundless depths of earth, the æther vast,
In vain the tongues of multitudes extol
Who see but little of the mighty all.

But others say the earth floats upon water. This view we consider the most ancient: it is ascribed to Thales the Milesian. It regards the earth as upheld in its place because it floats like a piece of wood or anything else of the same kind....But water itself cannot remain suspended on high, but must be upheld in its turn by something. Further, as air is lighter than water, so water is lighter than earth. How then can they fancy that what is lighter lies below and supports what is heavier? Again, were the whole earth able to float upon water, this would also be the case with its fragments. But this seems not so, for any piece of earth sinks to the bottom of water, and larger fragments sink more swiftly.

ARISTARCHUS AND ARCHIMEDES

Aristarchus of Samos, who flourished about 280 to 264 B.C., and Archimedes of Syracuse, born about 287 B.C., are the most modern in mind of the Greek physicists. Aristarchus alone directly concerns us here. The Pythagoreans had imagined a fire at the centre of the universe, but Aristarchus was the first to frame in a definite way the theory that the sun is the centre round which the earth and the other planets revolve. This does not appear in the only one of his works which survives, but it is made clear in the extract from Archimedes which follows later.

The application of mathematical reasoning to physics and astronomy, a method which has led to such tremendous results in modern times, is first seen in Aristarchus. The proofs, cast in geometrical form, are unsuited for our present purpose, but we illustrate the method by recording his hypotheses and his propositions.

ARISTARCHUS ON THE SIZES AND DISTANCES OF THE SUN AND MOON

(From *Aristarchus of Samos*, by Sir Thomas Heath.)

Hypotheses

1. That the moon receives its light from the sun.

2. That the earth is in the relation of a point and centre to the sphere in which the moon moves.

3. That, when the moon appears to us halved, the great

circle which divides the dark and the bright portions of the moon is in the direction of our eye.

4. That, when the moon appears to us halved, its distance from the sun is then less than a quadrant by one-thirtieth of a quadrant.

5. That the breadth of the (earth's) shadow is (that) of two moons.

6. That the moon subtends one-fifteenth part of a sign of the zodiac.

We are now in a position to prove the following propositions:

1. The distance of the sun from the earth is greater than eighteen times, but less than twenty times, the distance of the moon (from the earth); this follows from the hypothesis about the halved moon.

2. The diameter of the sun has the same ratio (as aforesaid) to the diameter of the moon.

3. The diameter of the sun has to the diameter of the earth a ratio greater than that which 19 has to 3, but less than that which 43 has to 6; this follows from the ratio thus discovered between the distances, the hypothesis about the shadow, and the hypothesis that the moon subtends one-fifteenth part of a sign of the zodiac.

ARCHIMEDES. THE SAND-RECKONER

(From *The Works of Archimedes*, Edited by Sir Thomas Heath.)

There are some, king Gelon, who think that the number of the sand is infinite in multitude; and I mean by the sand not only that which exists about Syracuse and the rest of Sicily but also that which is found in every region whether inhabited or uninhabited. Again there are some who, without regarding it as infinite, yet think that no number has been named which is great enough to exceed its multitude. And it is clear that they who hold this view, if they imagined a mass made up of sand in other respects as large as the mass of the earth, including in it all the seas and hollows of the earth filled up to a height equal to that of the highest of the mountains, would be many times further still from recognising that any number could be expressed which exceeded the multitude of the sand so taken. But

I will try to show you by means of geometrical proofs, which you will be able to follow, that, of the numbers named by me and given in the work which I sent to Zeuxippus, some exceed not only the number of the mass of sand equal in magnitude to the earth filled up in the way described, but also that of a mass equal in magnitude to the universe. Now you are aware that "universe" is the name given by most astronomers to the sphere whose centre is the centre of the earth and whose radius is equal to the straight line between the centre of the sun and the centre of the earth. This is the common account as you have heard from astronomers. But Aristarchus of Samos brought out a book consisting of some hypotheses, in which the premisses lead to the result that the universe is many times greater than that now so called. His hypotheses are that the fixed stars and the sun remain unmoved, that the earth revolves about the sun in the circumference of a circle, the sun lying in the middle of the orbit, and that the sphere of the fixed stars, situated about the same centre as the sun, is so great that the circle in which he supposes the earth to revolve bears such a proportion to the distance of the fixed stars as the centre of the sphere bears to its surface. Now it is easy to see that this is impossible; for, since the centre of the sphere has no magnitude, we cannot conceive it to bear any ratio whatever to the surface of the sphere. We must however take Aristarchus to mean this: since we conceive the earth to be, as it were, the centre of the universe, the ratio which the earth bears to what we describe as the "universe" is the same as the ratio which the sphere containing the circle in which he supposes the earth to revolve bears to the sphere of the fixed stars. For he adapts the proofs of his results to a hypothesis of this kind, and in particular he appears to suppose the magnitude of the sphere in which he represents the earth as moving to be equal to what we call the "universe."

I say then that, even if a sphere were made up of the sand, as great as Aristarchus supposes the sphere of the fixed stars to be, I shall still prove that, of the numbers named in the *Principles*, some exceed in multitude the number of the sand which is equal in magnitude to the sphere referred to....

COPERNICUS

The solar theory of Aristarchus did not commend itself to the astronomers who followed him. It is far more obvious to take the solid earth beneath us as the centre of the universe. Round it the celestial globe of the fixed stars is seen to revolve, and among those stars the sun and planets wander.

This view was developed mathematically about 130 B.C. by Hipparchus, the inventor of trigonometry. Hipparchus showed that the apparent motions could be explained by the supposition that the sun and planets moved round central points in orbits or epicycles, while these orbits were themselves carried round in larger orbits or cycles. Hipparchus' theory was expounded and recorded for us by Ptolemy of Alexandria, about 127–151 A.D., and held the field till the 15th century.

Rome never contributed much to original scientific thought, and, with the fall of Rome and the ruin of the Roman Empire, Alexandria remained the latest effective school of the ancient world. From Alexandria, as well as from the East, Arabian scholars helped to bring fragments of Greek learning to Western Europe after the Dark Ages, throughout which a few Latin commentaries formed the only direct link of knowledge.

In the revival of learning, a landmark was the recovery between 1210 and 1225 of the complete works of Aristotle, first rendered into Latin from imperfect Arabian versions, and then by direct translation from the Greek. The philosophy of Aristotle was welded into one with Christian dogma by Thomas Aquinas, and thus, when in the 15th and 16th centuries observation and experiment threw doubt on much of Aristotle's physical science, it was thought that religion was assailed also, and some ecclesiastical opposition was encountered.

Especially was this so, when Copernicus (1473–1543) revived the theory of astronomy which held the sun to be the centre of our system. Though published with the consent of Pope Clement VII, Copernicus' book was suspended in 1616 by the Vatican till it should be corrected by the omission of the heliocentric theory, which was declared "false and altogether opposed to Holy Scripture."

Nicolaus Koppernigk, who Latinized his name as Copernicus, was a Polish mathematician and astronomer. After a visit to Italy where the Pythagorean speculations were known, Copernicus began observations which led him to form a definite theory of the universe, taking the sun for its centre as in the scheme of Aristarchus. Copernicus showed that this theory explained the facts more simply than the cycles and epicycles of Hipparchus and Ptolemy. The first printed copy of his book reached him on his deathbed in 1543.

DE REVOLUTIONIBUS ORBIUM CELESTIUM

Lib. I. Cap. VIII.

The ancient philosophers affirm the earth to be at rest at the centre of the universe, and doubtless to maintain itself there. But, if anyone holds that the earth revolves, he affirms at any rate that the motion is natural and not forced. Which things indeed are according to nature, for motions which are forced produce different effects. It is true that things to which force or impact is applied necessarily disintegrate, and cannot hold together for long; but those which have a natural motion maintain themselves steadily, and are preserved in coherence. It is without reason, therefore, that Ptolemy fears lest the earth if it move should disintegrate and all terrestrial things be thrown into confusion by the power of nature, so far beyond that of art, or anything possible to human ingenuity. But why, on his view of a fixed earth and a moving sky, does he not fear it even more for the sky, whose motion is so much the more swift, as the heaven is greater than the earth? Is the sky perhaps an immense structure, which by the power of an ineffable motion is separated from its centre, and on the other hand falls in ruin if it stand still? Surely, if this be so, the magnitude of the heaven will increase to infinity. For by how much the further from the centre the motion is carried by its own impetus, by so much the motion will be swifter, on account of the ever-growing circumference which it must describe in order to traverse space in 24 hours: and, in turn, the immensity of the heaven grows greater by reason of this increasing motion. Thus the velocity increases the magnitude and the magnitude the velocity to infinity. And next comes the physical axiom: *That which is infinite is unable to travel forwards, nor can it move through any cause.* The heaven, therefore, necessarily stands still. But, they say, beyond the sky is neither body, nor space, nor vacuum—in a word nothing, and therefore no means exist of breaking out through the sky; then indeed it is wonderful, if from nothing anything is able to hold together. And if the heaven were infinite, and only bounded by its inner concavity, all the more can it be proved, perhaps, that there is nothing beyond the sky;

since each and everything is in it, whatever space it may occupy, the heaven will remain immoveable.... Whether therefore the universe be finite, or infinite, let us put aside the disputations of the physicists, having this certain—that the earth to its poles is bounded by a closed, spherical surface. Why, therefore, do

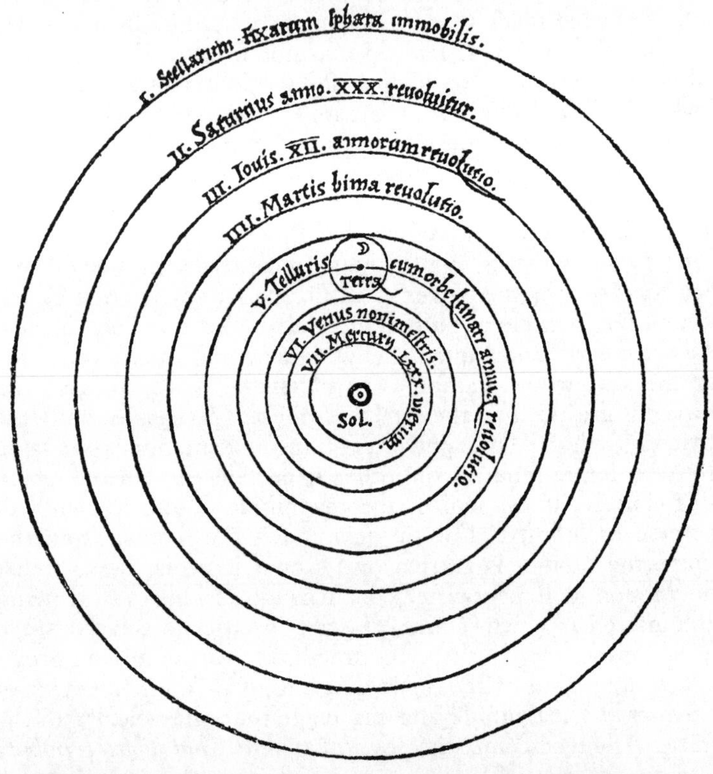

we hesitate to concede to it an appropriate mobility, rather than that the whole universe should fall in ruin, the end of which is unknown and unknowable; or confess of the daily revolution itself that in the heaven it is an appearance but in the earth reality? And this is indeed true, as Virgil's Aeneas affirms when he says

Provehimur portu, terræque urbesque recedunt.

CAP. X.

(Copernicus describes his idea of the universe.)

First and above all lies the sphere of the fixed stars, containing itself and all things, for that very reason immoveable; in truth the frame of the universe, to which the motion and position of all other stars are referred. Though some men think it to move in some way, we assign another reason why it appears to do so in our theory of the movement of the earth. Of the moving bodies first comes Saturn, who completes his circuit in xxx years. After him, Jupiter, moving in a twelve year revolution. Then Mars, who revolves biennially. Fourth in order an annual cycle takes place, in which we have said is contained the earth, with the lunar orbit as an epicycle. In the fifth place Venus is carried round in nine months. Then Mercury holds the sixth place, circulating in the space of eighty days. In the middle of all dwells the Sun. Who indeed in this most beautiful temple would place the torch in any other or better place than one whence it can illuminate the whole at the same time? Not ineptly, some call it the lamp of the universe, others its mind, others again its ruler—Trimegistus, the visible God, Sophocles' Electra the contemplation of all things. And thus rightly in as much as the Sun, sitting on a royal throne, governs the circumambient family of stars....We find, therefore, under this orderly arrangement, a wonderful symmetry in the universe, and a definite relation of harmony in the motion and magnitude of the orbs, of a kind it is not possible to obtain in any other way.

GALILEO GALILEI

THE heliocentric theory, thus revived by Copernicus after eighteen centuries' oblivion, gave a simpler explanation of the apparent movements of the sun and planets than did the theory of cycles and epicycles elaborated by Hipparchus and Ptolemy. For this reason alone its eventual victory was certain. But its real simplicity could not be appreciated till truer views of dynamics were obtained. While it was held that motion could only be maintained by the continual exertion of force, Aristotle's unmoved mover or some similar conception was required to explain the circulation of the planets.

The science of dynamics was brought into being and placed on the sound basis of experiment by Galileo Galilei (1564–1642), the descendant of a noble but impoverished house of Florence. Galileo, ceasing to reason as to how things ought to behave in the view of a philosopher, and refusing to take as final the teaching of Aristotle, made, first of the moderns, definite experiments on moving things. By dropping two bodies from the leaning tower of Pisa, he disproved Aristotle's dictum that heavy things fall faster than light, and, by rolling balls down inclined planes, he reduced the speed of fall till it could be measured easily, and discovered the law of constant acceleration in each second of time. A ball rolling down one plane ran up another of equal height whatever its inclination. Hence, if the second plane be made less and less steep, the ball travels farther and farther in virtue of the velocity acquired in its fall down the first plane. Finally, if the second plane be lowered till it becomes horizontal, it is clear from the experiments that the ball would travel forward for ever were it not slowly brought to rest by friction. Thus Galileo reached the idea of inertia—that a body tends to maintain its state, whether of rest or of uniform motion in a straight line. Aristotle's unmoved mover became unnecessary.

But, if Galileo's dynamics constitute his chief claim to scientific fame, his invention of the telescope and the astronomical observations he made with it caused a greater stir at the time. In particular, his discovery of the satellites of Jupiter demonstrated the existence of an arrangement which, on a smaller scale, illustrated the Copernican theory of the solar system, and the irregularities of the surface of the moon, shown clearly through his telescope, first gave convincing proof of the imperfection of the heavenly bodies, and prepared men's minds for the fact that the celestial orbs, to Aristotle divine and incorruptible, were of like nature with terrestrial objects, and subject to the same processes of motion, change and decay.

(*The Sidereal Messenger* of Galileo Galilei, translated by Edward Stafford Carlos, M.A., 1880.)

The

SIDEREAL MESSENGER

unfolding great and marvellous sights,
and proposing them to the attention of everyone,
but especially philosophers and astronomers,

being such as have been observed by

GALILEO GALILEI

a gentleman of Florence,
Professor of Mathematics in the University of Padua,

with the aid of a

TELESCOPE

lately invented by him,

Respecting the Moon's Surface, an innumerable number of Fixed Stars, the Milky Way, and Nebulous Stars, but especially respecting Four Planets which revolve round the Planet Jupiter at different distances and in different periodic times, with amazing velocity, and which, after remaining unknown to every one up to this day, the Author recently discovered, and determined to name the

MEDICEAN STARS.

Venice 1610.

Introduction

In the present small treatise I set forth some matters of great interest for all observers of natural phenomena to look at and consider. They are of great interest, I think, first, from their intrinsic excellence; secondly, from their absolute novelty; and lastly, also on account of the instrument by the aid of which they have been presented to my apprehension.

The number of the Fixed Stars which observers have been able to see without artificial powers of sight up to this day can be counted. It is therefore decidedly a great feat to add to their number, and to set distinctly before the eyes other stars in myriads, which have never been seen before, and which surpass the old, previously known, stars in number more than ten times.

Again, it is a most beautiful and delightful sight to behold the body of the Moon, which is distant from us nearly sixty semi-diameters of the Earth, as near as if it was at a distance of only two of the same measures; so that the diameter of this same Moon appears about thirty times larger, its surface about nine hundred times, and its solid mass nearly 27,000 times larger than when it is viewed only with the naked eye: and consequently any one may know with the certainty that is due to the use of our senses, that the Moon certainly does not possess a smooth and polished surface, but one rough and uneven, and, just like the face of the Earth itself, is everywhere full of vast protuberances, deep chasms, and sinuosities.

Then to have got rid of disputes about the Galaxy or Milky Way, and to have made its nature clear to the very senses, not to say to the understanding, seems by no means a matter which ought to be considered of slight importance. In addition to this, to point out, as with one's finger, the nature of those stars which every one of the astronomers up to this time has called *nebulous*, and to demonstrate that it is very different from what has hitherto been believed, will be pleasant, and very fine. But that which will excite the greatest astonishment by far, and which indeed especially moved me to call the attention of all astronomers and philosophers, is this, namely, that I have discovered four planets, neither known nor observed by any one of the astronomers before my time, which have their orbits round a certain bright star, one of those previously known, like Venus and Mercury round the Sun, and are sometimes in front of it, sometimes behind it, though they never depart from it beyond certain limits. All which facts were discovered and observed a few days ago by the help of a telescope devised by me, through God's grace first enlightening my mind.

Perchance other discoveries still more excellent will be made from time to time by me or other observers, with the assistance of a similar instrument, so I will first briefly record its shape and preparation, as well as the occasion of its being devised, and then I will give an account of the observations made by me.

GALILEO'S *account of the* INVENTION *of his* TELESCOPE

About ten months ago a report reached my ears that a Dutchman had constructed a telescope, by the aid of which visible objects, although at a great distance from the eye of the observer, were seen distinctly as if near; and some proofs of its most wonderful performances were reported, which some gave credence to, but others contradicted. A few days after, I received confirmation of the report in a letter written from Paris by a noble Frenchman, Jaques Badovere, which finally determined me to give myself up first to inquire into the principle of the telescope, and then to consider the means by which I might compass the invention of a similar instrument, which after a little while I succeeded in doing, through deep study of the theory of Refraction; and I prepared a tube, at first of lead, in the ends of which I fitted two glass lenses, both plane on one side, but on the other side one spherically convex, and the other concave. Then bringing my eye to the concave lens I saw objects satisfactorily large and near, for they appeared one-third of the distance off and nine times larger than when they are seen with the natural eye alone. I shortly afterwards constructed another telescope with more nicety, which magnified objects more than sixty times. At length, by sparing neither labour nor expense, I succeeded in constructing for myself an instrument so superior that objects seen through it appear magnified nearly a thousand times, and more than thirty times nearer than if viewed by the natural powers of sight alone.

GALILEO'S *first* OBSERVATIONS *with his* TELESCOPE

It would be altogether a waste of time to enumerate the number and importance of the benefits which this instrument may be expected to confer, when used by land or sea. But without paying attention to its use for terrestrial objects, I betook myself to observations of the heavenly bodies; and first of all, I viewed the Moon as near as if it was scarcely two semi-diameters of the Earth distant. After the Moon, I frequently observed other heavenly bodies, both fixed stars and planets, with incredible delight; and, when I saw their very great number, I began to consider about a method by which I might be able to measure their distances apart, and at length I found one. And here it is fitting that all who intend to turn their attention to observations of this kind should receive certain cautions. For, in the first place, it is absolutely necessary for them to prepare a most perfect telescope, one which will show very bright objects distinct and free from any mistiness, and will magnify them at least 400 times, for then it will show them as if only one-twentieth of their distance off. For unless the instrument be of such power, it will be in vain to attempt to view all the things which have been seen by me in the heavens, or which will be enumerated hereafter.

Method of determining the MAGNIFYING POWER *of the* TELESCOPE

But in order that any one may be a little more certain about the magnifying power of his instrument, he shall fashion two circles, or two square pieces of paper, one of which is 400 times greater than the other, but that will be when the diameter of the greater is twenty times the length of the diameter of the other. Then he shall view from a distance simultaneously both surfaces, fixed on the same wall, the smaller with one eye applied to the telescope, and the larger with the other eye unassisted: for that may be done without inconvenience at one and the same instant with both eyes open. Then both figures will appear of the same size, if the instrument magnifies objects in the desired proportion.

After such an instrument has been prepared, the method of measuring distances remains for inquiry, and this we shall accomplish by the following contrivance:

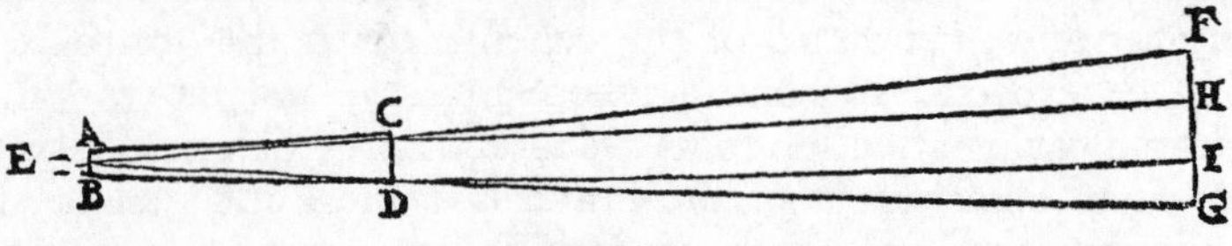

Fig. 1.

Method of measuring small angular distances between HEAVENLY BODIES *by the* SIZE *of the* APERTURE *of the* TELESCOPE

For the sake of being more easily understood, I will suppose a tube *ABCD*. Let *E* be the eye of the observer; then, when there are no lenses in the tube rays from the eye to the object *FG* would be drawn in the straight lines *ECF*, *EDG*, but when the lenses have been inserted, let the rays go in the bent lines *ECH*, *EDI*,—for they are contracted, and those which originally, when unaffected by the lenses, were directed to the object *FG*, will include only the part *HI*. Hence the ratio of the distance *EH* to the line *HI* being known, we shall be able to find, by means of a table of sines, the magnitude of the angle subtended at the eye by the object *HI*, which we shall find to contain only some minutes. But if we fit on the lens *CD* thin plates of metal, pierced, some with larger, others with smaller apertures, by putting on over the lens sometimes one plate, sometimes another, as may be necessary, we shall construct at our pleasure different subtending angles of more or fewer minutes, by the help of which we shall be able to measure conveniently the intervals between stars separated by an angular distance of some minutes, within an error of one or two minutes. But let it suffice for the present to have thus slightly touched, and as it were just put our lips to these matters, for on some other opportunity I will publish the theory of this instrument in completeness.

Now let me review the observations made by me during the two months just past, again inviting the attention of all who are eager for true philosophy to the beginnings which led to the sight of most important phenomena.

The MOON. *Ruggedness of its surface. Existence of* LUNAR MOUNTAINS *and* VALLEYS

Let me first speak of the surface of the Moon, which is turned towards us. For the sake of being understood more easily, I distinguish two parts in it, which I call respectively the brighter and the darker. The brighter part seems to surround and pervade the whole hemisphere; but the darker part, like a sort of cloud, discolours the Moon's surface and makes it appear covered with spots. Now these spots, as they are somewhat dark and of considerable size, are plain to every one, and every age has seen them, wherefore I shall call them *great* or *ancient* spots, to distinguish them from other spots, smaller in size, but so thickly scattered that they sprinkle the whole surface of the Moon, but especially the brighter portion of it. These spots have never been observed by any one before me; and from my observations of them, often repeated, I have been led to that opinion which I have expressed, namely, that I feel sure that the surface of the Moon is not perfectly smooth, free from inequalities and exactly spherical, as a large school of philosophers considers with regard to the Moon and the other heavenly bodies, but that, on the contrary, it is full of inequalities, uneven, full of hollows and protuberances, just like the surface of the Earth itself, which is varied everywhere by lofty mountains and deep valleys.

The appearances from which we may gather these conclusions are of the following nature:—On the fourth or fifth day after new-moon, when the Moon presents itself to us with bright horns, the boundary which divides the part in shadow from the enlightened part does not extend continuously in an ellipse, as would happen in the case of a perfectly spherical body, but it is marked out by an irregular, uneven, and very wavy line...for several bright excrescences, as they may be called, extend beyond the boundary of light and shadow into the dark part, and on the other hand pieces of shadow encroach upon the light—nay, even a great quantity of small blackish spots, altogether separated from the dark part, sprinkle everywhere almost the whole space which is at the time flooded with the Sun's light, with the exception of that part alone which is occupied by the great and ancient spots. I have noticed that the small spots just mentioned have this common characteristic always and in every case, that they have the dark part towards the Sun's position, and on the side away from the Sun they have brighter boundaries, as if they were crowned with shining summits. Now we have an appearance quite similar on the Earth about sunrise, when we behold the valleys, not yet flooded with light, but the mountains surrounding them on the side opposite to the Sun already ablaze with the splendour of his beams; and just as the shadows in the hollows of the Earth diminish in size as the Sun rises higher, so also these spots on the Moon lose their blackness as the illuminated part grows larger and larger. Again, not only are the boundaries of light and shadow in the Moon seen to be uneven and sinuous, but—and this produces still greater astonishment—there appear very many bright points within the darkened portion of the Moon, altogether divided and broken off from the illuminated tract, and separated from it by no inconsiderable interval, which, after a little while, gradually increase in size and brightness, and after an hour or two become joined on to the rest of the main portion, now become somewhat larger; but in the meantime others, one here and another there, shooting up as if growing, are lighted up within the shaded portion, increase in size, and at last are linked on to the same luminous surface, now still more extended. ...Now, is it not the case on the Earth before sunrise, that while the level plain is still in shadow, the peaks of the most lofty mountains are illuminated by the Sun's rays? After a little while does not the light spread further, while the middle and larger parts of those mountains are becoming illuminated; and at length, when the Sun has risen, do not the illuminated parts of the plains and hills join together? The grandeur, however, of such prominences and depressions in the Moon seems to surpass both in magnitude and extent the ruggedness of the Earth's surface, as I shall hereafter show....

Stars. *Their* appearance *in the* telescope

Hitherto I have spoken of the observations which I have made concerning the Moon's body; now I will briefly announce the phenomena which have been, as yet, seen by me with reference to the Fixed Stars. And first of all the following fact is worthy of consideration:—The stars, fixed as well as erratic, when seen with a telescope, by no means appear to be increased in magnitude in the same proportion as other objects, and the Moon herself, gain increase of size; but in the case of the stars such an increase appears much less, so that you may consider that a telescope, which (for the sake of illustration) is powerful enough to magnify other objects a hundred times, will scarcely render the stars magnified four or five times. But the reason of this is as follows: When stars are viewed with our natural eyesight they do not present themselves to us of their bare, real size, but beaming with a certain vividness, and fringed with sparkling rays, especially when the night is far advanced; and from this circumstance they appear much larger than they would if they were stripped of those adventitious fringes, for the angle which they subtend at the eye is determined not by the primary disc of the star, but by the brightness which so widely surrounds it....A telescope... removes from the stars their adventitious and accidental splendours before it enlarges their true discs (if indeed they are of that shape), and so they seem less magnified than other objects, for a star of the fifth or sixth magnitude seen through a telescope is shown as of the first magnitude only.

The difference between the appearance of the planets and the fixed stars seems also deserving of notice. The planets present their discs perfectly round, just as if described with a pair of compasses, and appear as so many little moons, completely illuminated and of a globular shape; but the fixed stars do not look to the naked eye bounded by a circular circumference, but rather like blazes of light, shooting out beams on all sides and very sparkling, and with a telescope they appear of the same shape as when they are viewed by simply looking at them, but so much larger that a star of the fifth or sixth magnitude seems to equal Sirius, the largest of all the fixed stars.

TELESCOPIC STARS: *their infinite multitude. As examples,* ORION'S BELT *and* SWORD *and the* PLEIADES *are described as seen by* GALILEO

But beyond the stars of the sixth magnitude you will behold through the telescope a host of other stars, which escape the unassisted sight, so numerous as to be almost beyond belief, for you may see more than six other differences of magnitude, and the largest of these, which I may call stars of the seventh magnitude, or of the first magnitude of invisible stars, appear with the aid of the telescope larger and brighter than stars of the second magnitude seen with the unassisted sight. But in order that you may see one or two proofs of the inconceivable manner in which they are crowded together, I have determined to make out a case against two star-clusters, that from them as a specimen you may decide about the rest.

As my first example I had determined to depict the entire constellation of Orion, but I was overwhelmed by the vast quantity of stars and by want of time, and so I have deferred attempting this to another occasion, for there are adjacent to, or scattered among, the old stars more than five hundred new stars within the limits of one or two degrees. For this reason I have selected the three stars in Orion's Belt and the six in his

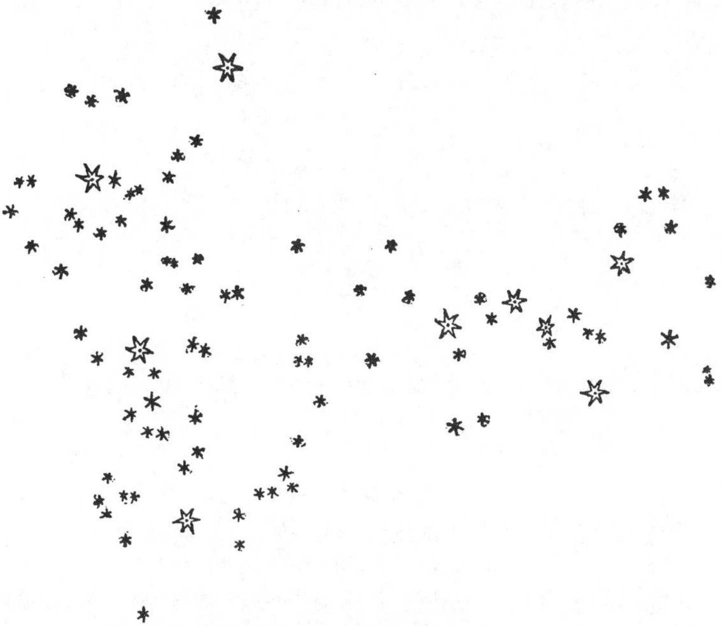

Fig. 2. Orion's Belt and Sword.

Sword, which have been long well-known groups, and I have added eighty other stars recently discovered in their vicinity, and I have preserved as exactly as possible the intervals between them. The well-known or old stars, for the sake of distinction, I have depicted of larger size, and I have outlined them with a double line; the others, invisible to the naked eye, I have marked smaller and with one line only. I have also preserved the differences of magnitude as much as I could. As a second example I have depicted the six stars of the constellation Taurus, called the

Pleiades (I say *six* intentionally, since the seventh is scarcely ever visible), a group of stars which is enclosed in the heavens within very narrow precincts. Near these there lie more than forty others invisible to the naked eye, no one of which is more than half a degree off any of the aforesaid six; of these I have noticed only thirty-six in my diagram. I have preserved their intervals, magnitudes, and the distinction between the old and the new stars, just as in the case of the constellation Orion.

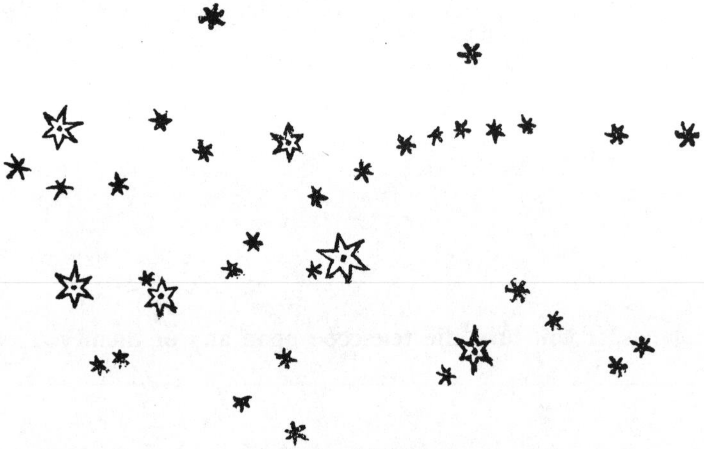

Fig. 3. Pleiades.

The MILKY WAY *consists entirely of* STARS *in countless numbers and of various magnitudes*

The next object which I have observed is the essence or substance of the Milky Way. By the aid of a telescope any one may behold this in a manner which so distinctly appeals to the senses that all the disputes which have tormented philosophers through so many ages are exploded at once by the irrefragable evidence of our eyes, and we are freed from wordy disputes upon this subject, for the Galaxy is nothing else but a mass of innumerable stars planted together in clusters. Upon whatever part of it you direct the telescope straightway a vast crowd of stars presents itself to view; many of them are tolerably large and extremely bright, but the number of small ones is quite beyond determination.

NEBULÆ, *resolved into clusters of stars: as examples, the* NEBULA *in* ORION'S HEAD *and* PRÆSEPE

And whereas that milky brightness, like the brightness of a white cloud, is not only to be seen in the Milky Way, but several

Fig. 4. Orion's Head.

spots of a similar colour shine faintly here and there in the heavens, if you turn the telescope upon any of them you will

Fig. 5. Præsepe.

find a cluster of stars packed close together. Further—and you will be more surprised at this—the stars which have been called by every one of the astronomers up to this day *nebulous*, are groups of small stars set thick together in a wonderful way, and although each one of them on account of its smallness, or its immense distance from us, escapes our sight, from the commingling of their rays there arises that brightness which has hitherto been believed to be the denser part of the heavens, able to reflect the rays of the stars or the Sun.

I have observed some of these, and I wish to subjoin the star-clusters of two of these nebulæ. First, you have a diagram of the nebula called that of Orion's Head, in which I have counted twenty-one stars.

The second cluster contains the nebula called Præsepe, which is not one star only, but a mass of more than forty small stars. I have noticed thirty-six stars, besides the Aselli, arranged in the order of the accompanying diagram.

DISCOVERY *of* JUPITER'S SATELLITES, *Jan.* 7, 1610*: record of* GALILEO'S OBSERVATIONS *during* TWO MONTHS

I have now finished my brief account of the observations which I have thus far made with regard to the Moon, the Fixed Stars, and the Galaxy. There remains the matter, which seems to me to deserve to be considered the most important in this work, namely, that I should disclose and publish to the world the occasion of discovering and observing four PLANETS, never seen from the very beginning of the world up to our own times, their positions, and the observations made during the last two months about their movements and their changes of magnitude; and I summon all astronomers to apply themselves to examine and determine their periodic times, which it has not been permitted me to achieve up to this day, owing to the restriction of my time. I give them warning however again, so that they may not approach such an inquiry to no purpose, that they will want a very accurate telescope, and such as I have described in the beginning of this account.

On the 7th day of January in the present year, 1610, in the first[1] hour of the following night, when I was viewing the

[1] The times of Galileo's observations are to be understood as reckoned from sunset.

constellations of the heavens through a telescope, the planet Jupiter presented itself to my view, and as I had prepared for myself a very excellent instrument, I noticed a circumstance which I had never been able to notice before, owing to want of power in my other telescope, namely, that three little stars, small but very bright, were near the planet; and although I believed them to belong to the number of the fixed stars, yet they made me somewhat wonder, because they seemed to be arranged exactly in a straight line, parallel to the ecliptic, and to be brighter than the rest of the stars, equal to them in magnitude. The position of them with reference to one another and to Jupiter was as follows:

Ori. * * O * Occ.

Fig. 1.

On the east side there were two stars, and a single one towards the west. The star which was furthest towards the east, and the western star, appeared rather larger than the third.

I scarcely troubled at all about the distance between them and Jupiter, for, as I have already said, at first I believed them to be fixed stars; but when on January 8th, led by some fatality, I turned again to look at the same part of the heavens, I found a very different state of things, for there were three little stars all west of Jupiter, and nearer together than on the previous night, and they were separated from one another by equal intervals, as the accompanying figure shows.

Ori. O * * * Occ.

Fig. 2.

At this point, although I had not turned my thoughts at all upon the approximation of the stars to one another, yet my surprise began to be excited, how Jupiter could one day be found to the east of all the aforesaid fixed stars when the day before it had been west of two of them; and forthwith I became afraid lest the planet might have moved differently from the calculation of astronomers, and so had passed those stars by its own proper motion. I therefore waited for the next night with the most intense longing, but I was disappointed of my hope, for the sky was covered with clouds in every direction.

But on January 10th the stars appeared in the following position with regard to Jupiter, the third, as I thought, being

Ori. * * O Occ.

Fig. 3.

hidden by the planet. They were situated just as before, exactly in the same straight line with Jupiter, and along the Zodiac.

When I had seen these phenomena, as I knew that corresponding changes of position could not by any means belong to Jupiter, and as, moreover, I perceived that the stars which I saw had always been the same, for there were no others either in front or behind, within a great distance, along the Zodiac,—at length, changing from doubt into surprise, I discovered that the interchange of position which I saw belonged not to Jupiter, but to the stars to which my attention had been drawn, and I thought therefore that they ought to be observed henceforward with more attention and precision.

Accordingly, on January 11th I saw an arrangement of the following kind:

Ori. * * O Occ.

Fig. 4.

namely, only two stars to the east of Jupiter, the nearer of which was distant from Jupiter three times as far as from the star further to the east; and the star furthest to the east was nearly twice as large as the other one; whereas on the previous night they had appeared nearly of equal magnitude. I therefore concluded, and decided unhesitatingly, that there are three stars in the heavens moving about Jupiter, as Venus and Mercury round the Sun; which at length was established as clear as daylight by numerous other subsequent observations. These observations also established that there are not only three, but four, erratic sidereal bodies performing their revolutions round Jupiter, observations of whose changes of position made with more exactness on succeeding nights the following account will supply. I have measured also the intervals between them with the telescope in the manner already explained. Besides this, I have given the times of observation, especially when several were made in the same night, for the revolutions of these planets are so swift that an observer may generally get differences of position every hour.

Jan. 12.—At the first hour of the next night I saw these heavenly bodies arranged in this manner:

Ori. * *○ * Occ.

Fig. 5.

The satellite furthest to the east was greater than the satellite furthest to the west; but both were very conspicuous and bright; the distance of each one from Jupiter was two minutes. A third satellite, certainly not in view before, began to appear at the third hour: it nearly touched Jupiter on the east side, and was exceedingly small. They were all arranged in the same straight line, along the ecliptic.

Jan. 13.—For the first time four satellites were in view in the following position with regard to Jupiter.

Ori. * ○ * * * Occ.

Fig. 6.

There were three to the west, and one to the east; they made a straight line nearly, but the middle satellite of those to the west deviated a little from the straight line towards the north. The satellite furthest to the east was at a distance of 2′ from Jupiter; there were intervals of 1′ only between Jupiter and the nearest satellite, and between the satellites themselves, west of Jupiter. All the satellites appeared of the same size, and though small they were very brilliant, and far outshone the fixed stars of the same magnitude.

Jan. 14.—The weather was cloudy.

Etc....

These are my observations upon the four Medicean planets, recently discovered for the first time by me; and although it is not yet permitted me to deduce by calculation from these observations the orbits of these bodies, yet I may be allowed to make some statements, based upon them, well worthy of attention.

DEDUCTIONS *from the previous observations concerning the orbits and periods of* JUPITER'S SATELLITES

And, in the first place, since they are sometimes behind, sometimes before Jupiter, at like distances, and withdraw from this planet towards the east and towards the west only within very narrow limits of divergence, and since they accompany this planet alike when its motion is retrograde and direct, it can be a matter of doubt to no one that they perform their revolutions about this planet, while at the same time they all accomplish together orbits of twelve years' length about the centre of the world. Moreover, they revolve in unequal circles, which is evidently the conclusion to be drawn from the fact that I have never been permitted to see two satellites in conjunction when their distance from Jupiter was great, whereas near Jupiter two, three, and sometimes all four, have been found closely packed together. Moreover, it may be detected that the revolutions of the satellites which describe the smallest circles round Jupiter are the most rapid, for the satellites nearest to Jupiter are often to be seen in the east, when the day before they have appeared in the west, and contrariwise. Also the satellite moving in the greatest orbit seems to me, after carefully weighing the occasions of its returning to positions previously noticed, to have a periodic time of half a month. Besides, we have a notable and splendid argument to remove the scruples of those who can tolerate the revolution of the planets round the Sun in the Copernican system, yet are so disturbed by the motion of one Moon about the Earth, while both accomplish an orbit of a year's length about the Sun, that they consider that this theory of the universe must be upset as impossible: for now we have not one planet only revolving about another, while both traverse a vast orbit about the Sun, but our sense of sight presents to us four satellites circling about Jupiter, like the Moon about the Earth, while the whole system travels over a mighty orbit about the Sun in the space of twelve years.

NEWTON

In the year 1642 Galileo died and Newton was born.

Galileo had prepared the way by demonstrating the properties of moving bodies, and by inventing the telescope, which made possible accurate astronomical measurement. The possession of one of Galileo's instruments enabled his contemporary John Kepler to extend the observations of his master Tycho Brahe, and finally to summarize all the varied movements of the planets by simple laws. Of these the most important was that the planets moved in elliptical orbits round the sun at one of the two foci.

By Galileo's principle of inertia it was clear that a planet would continue in motion if unobstructed, but the reason for its deflection from a straight path into a closed orbit round the sun was yet to seek. The idea of an attracting force was perhaps in the air, and Newton, turning his unrivalled mathematical genius to the problem, proved that, if the force were inversely proportional to the square of the distance, the path would be an ellipse round the attracting body in one of the foci, in conformity with Kepler's observed law for the planets. Moreover, the familiar phenomenon of weight suggested an attraction by the earth, measured by the rate at which bodies fell to the ground. Why should not this be the same force which kept the moon in her orbit? Reduced by the inverse square law, its magnitude at the distance of the moon could be calculated. Misled by a false estimate of the moon's distance, at first Newton found a discrepancy, and put away his calculations. But hearing of a newer estimate, he took them up again, and in excitement, it is said, so great he could scarce see his figures, he found conformity between his theory and the facts. The moon was but a stone ever falling towards the earth, and so kept in its orbit. The terrestrial and familiar force of gravity controlled the path of a heavenly body.

THE MATHEMATICAL PRINCIPLES OF NATURAL PHILOSOPHY

By Sir Isaac Newton.

London: 1687.

Translated into English by Andrew Motte.

To which are added, *Newton's System of the World*; etc.

London: 1803

From the Author's Preface

Since the antients (as we are told by *Pappus*) made great account of the science of mechanics in the investigation of natural things;

and the moderns, laying aside substantial forms and occult qualities, have endeavoured to subject the phænomena of nature to the laws of mathematics, I have in this treatise cultivated mathematics so far as it regards philosophy....Our design not respecting arts, but philosophy, and our subject not manual but natural powers, we consider chiefly those things which relate to gravity, levity, elastic force, the resistance of fluids, and the like forces, whether attractive or impulsive; and therefore we offer this work as mathematical principles of philosophy; for all the difficulty of philosophy seems to consist in this—from the phænomena of motions to investigate the forces of nature, and then from these forces to demonstrate the other phænomena; and to this end the general propositions in the first and second book are directed. In the third book we give an example of this in the explication of the System of the World; for by the propositions mathematically demonstrated in the first book, we there derive from the celestial phænomena the forces of gravity with which bodies tend to the sun and the several planets. Then from these forces, by other propositions which are also mathematical, we deduce the motions of the planets, the comets, the moon, and the sea. I wish we could derive the rest of the phænomena of nature by the same kind of reasoning from mechanical principles; for I am induced by many reasons to suspect that they may all depend upon certain forces by which the particles of bodies, by some causes hitherto unknown, are either mutually impelled towards each other, and cohere in regular figures, or are repelled and recede from each other; which forces being unknown, philosophers have hitherto attempted the search of nature in vain; but I hope the principles here laid down will afford some light either to that or some truer method of philosophy.

In the publication of this work the most acute and universally learned Mr *Edmund Halley* not only assisted me with his pains in correcting the press and taking care of the schemes; but it was owing to his solicitations that its becoming public is owing; for when he had obtained of me my demonstrations of the figure of the celestial orbits, he continually pressed me to communicate the same to the *Royal Society*, who afterwards, by their kind encouragement and entreaties, engaged me to think of publishing them. But after I had begun to consider the inequalities of the lunar motions, and had entered upon some other things relating to the laws and measures of gravity, and other forces; and the figures that would be described by bodies attracted according to given laws; and the motion of several bodies moving among themselves; the motion of bodies in resisting mediums; the forces, densities, and motions, of mediums; the orbits of the comets, and such like; I put off that publication till I had made a search into those matters, and could put out the whole together. What relates to the lunar motions (being imperfect) I have put all together in the corollaries of prop. 66, to avoid being obliged to propose and distinctly demonstrate the several things there contained in a method more prolix than the subject deserved, and interrupt the series of the several propositions. Some things, found out after the rest, I chose to insert in places less suitable, rather than change the number of the propositions and the citations. I heartily beg that what I have here done may be read with candour; and that the defects I have been guilty of upon this difficult subject may be not so much reprehended as kindly supplied, and investigated by new endeavours of my readers.

ISAAC NEWTON.

CAMBRIDGE, TRINITY COLLEGE,
May 8, 1686.

From THE SYSTEM OF THE WORLD

IT was the antient opinion of not a few, in the earliest ages of philosophy, that the fixed stars stood immoveable in the highest parts of the world; that under the fixed stars the planets were carried about the sun; that the earth, as one of the planets, described an annual course about the sun, while by a diurnal motion it was in the mean time revolved about its own axis; and that the sun, as the common fire which served to warm the whole, was fixed in the centre of the universe....

It is not to be denied but that *Anaxagoras*, *Democritus*, and others, did now and then start up, who would have it that the earth possessed the centre of the world, and that the stars of all sorts were revolved towards the west about the earth quiescent in the centre, some at a swifter, others at a slower rate.

However, it was agreed on both sides that the motions of the celestial bodies were performed in spaces altogether free and void of resistance. The whim of solid orbs was of a later date, introduced by *Eudoxus*, *Calippus* and *Aristotle*; when the antient philosophy began to decline, and to give place to the new prevailing fictions of the Greeks....

Whence it was that the planets came to be retained within any certain bounds in these free spaces, and to be drawn off from the rectilinear courses, which, left to themselves, they should have pursued, into regular revolutions in curvilinear orbits, are questions which we do not know how the antients explained; and probably it was to give some sort of satisfaction to this difficulty that solid orbs were introduced.

The later philosophers pretend to account for it either by the action of certain vortices, as *Kepler* and *Des Cartes*; or by some other principle of impulse or attraction, as *Borelli*, *Hooke*, and others of our nation; for, from the laws of motion, it is most certain that these effects must proceed from the action of some force or other.

But our purpose is only to trace out the quantity and properties of this force from the phænomena, and to apply what we discover in some simple cases as principles, by which, in a mathematical way, we may estimate the effects thereof in more involved cases; for it would be endless and impossible to bring every particular to direct and immediate observation.

We said, *in a mathematical way*, to avoid all questions about the nature or quality of this force, which we would not be understood to determine by any hypothesis; and therefore call it by the general name of a centripetal force, as it is a force which is directed towards some centre; and as it regards more particularly a body in that centre, we call it circum-solar, circum-terrestrial, circum-jovial; and in like manner in respect of other central bodies....

That there are centripetal forces actually directed to the bodies of the sun, of the earth, and other planets, I thus infer.

The moon revolves about our earth, and by radii drawn to its centre describes areas nearly proportional to the times in which they are described, as is evident from its velocity compared with its apparent diameter; for its motion is slower when its diameter is less (and therefore its distance greater), and its motion is swifter when its diameter is greater.

The revolutions of the satellites of Jupiter about that planet are more regular; for they describe circles concentric with Jupiter by equable motions, as exactly as our senses can distinguish.

And so the satellites of Saturn are revolved about this planet with motions nearly circular and equable, scarcely disturbed by any eccentricity hitherto observed.

That Venus and Mercury are revolved about the sun, is demonstrable from their moon-like appearances: when they shine with a full face, they are in those parts of their orbs which in respect of the earth lie beyond the sun; when they appear half full, they are in those parts which lie over against the sun; when horned, in those parts which lie between the earth and the sun; and sometimes they pass over the sun's disk, when directly interposed between the earth and the sun.

And Venus, with a motion almost uniform, describes an orb nearly circular and concentric with the sun.

But Mercury, with a more eccentric motion, makes remarkable approaches to the sun, and goes off again by turns; but it is always swifter as it is near to the sun, and therefore by a radius drawn to the sun still describes areas proportional to the times.

Lastly, that the earth describes about the sun, or the sun about the earth, by a radius from the one to the other, areas exactly proportional to the times, is demonstrable from the apparent diameter of the sun compared with its apparent motion.

These are astronomical experiments; from which it follows, by prop. 1, 2, 3, in the first book of our *Principles*, and their corollaries, that there are centripetal forces actually directed (either accurately or without considerable error) to the centres of the earth, of Jupiter, of Saturn, and of the sun. In Mercury, Venus, Mars, and the lesser planets, where experiments are wanting, the arguments from analogy must be allowed in their place.

That those forces decrease in the duplicate proportion of the distances from the centre of every planet, appears by cor. 6, prop. 4, book 1; for the periodic times of the satellites of Jupiter are one to another in the sesquiplicate proportion of their distances from the centre of this planet.

This proportion has been long ago observed in those satellites;

and Mr. *Flamsted*, who had often measured their distances from Jupiter by the micrometer, and by the eclipses of the satellites, wrote to me, that it holds to all the accuracy that possibly can be discerned by our senses. And he sent me the dimensions of their orbits taken by the micrometer, and reduced to the mean distance of Jupiter from the earth, or from the sun, together with the times of their revolutions, as follows:

The greatest elongation of the satellites from the centre of Jupiter as seen from the sun.	The periodic times of their revolutions.
	d. h.
1st ... 1′ 48″ or 108″	1 18 28′ 36″
2nd ... 3 01 or 181	3 13 17 54
3rd ... 4 46 or 286	7 03 59 36
4th ... 8 $13\frac{1}{2}$ or $493\frac{1}{2}$	16 18 5 13

Whence the sesquiplicate proportion may be easily seen. For example: the $16^{d.}\ 18^{h.}\ 05'\ 13''$ is to the time $1^{d.}\ 18^{h.}\ 28'\ 36''$ as $493\frac{1}{2}'' \times \sqrt{493\frac{1}{2}''}$ to $108'' \times \sqrt{108''}$, neglecting those small fractions which, in observing, cannot be certainly determined.

Before the invention of the micrometer, the same distances were determined in semi-diameters of Jupiter thus:

Distance of the 1st.	2d.	3d.	4th.
By *Galileo* 6	10	16	28
Simon Marius.........6	10	16	26
Cassini 5	8	13	23
Borelli more exactly ......$5\frac{2}{3}$	$8\frac{2}{3}$	14	$24\frac{2}{3}$

After the invention of the micrometer.

By *Townley* 5,51	8,78	13,47	24,72
Flamsted 5,31	8,85	13,98	24,23
More accurately by the eclipses ... 5,578	8,876	14,159	24,903

And the periodic times of those satellites, by the observations of Mr. *Flamsted*, are

$1^{d.}\ 18^{h.}\ 28'\ 36'' \mid 3^{d.}\ 17^{h.}\ 17'\ 54''$
$7^{d.}\ 3^{h.}\ 59'\ 36'' \mid 16^{d.}\ 18^{h.}\ 5'\ 13''$,

as above.

And the distances thence computed are

5,578 | 8,878 | 14,168 | 24,968,

accurately agreeing with the distances by observation.

Cassini assures us (p. 164, 165) that the same proportion is observed in the circum-saturnal planets. But a longer course of observations is required before we can have a certain and accurate theory of those planets.

In the circum-solar planets, Mercury and Venus, the same proportion holds with great accuracy, according to the dimensions of their orbs, as determined by the observations of the best astronomers.

That Mars is revolved about the sun is demonstrated from the phases which it shows, and the proportion of its apparent diameters; for from its appearing full near conjunction with the sun, and gibbous in its quadratures, it is certain that it surrounds the sun.

And since its apparent diameter appears about five times greater when in opposition to the sun than when in conjunction therewith, and its distance from the earth is reciprocally as its apparent diameter, that distance will be about five times less when in opposition to than when in conjunction with the sun: but in both cases its distance from the sun will be nearly about the same with the distance which is inferred from its gibbous appearance in the quadratures. And as it encompasses the sun at almost equal distances, but in respect of the earth is very unequally distant, so by radii drawn to the sun it describes areas nearly uniform; but by radii drawn to the earth, it is sometimes swift, sometimes stationary, and sometimes retrograde....

Kepler and *Bullialdus* have, with great care, determined the distances of the planets from the sun: and hence it is that their tables agree best with the heavens. And in all the planets, in Jupiter and Mars, in Saturn and the Earth, as well as in Venus and Mercury, the cubes of their distances are as the squares of their periodic times; and therefore (by cor. 6, prop. 4) the centripetal circum-solar force throughout all the planetary regions decreases in the duplicate proportion of the distances from the sun. In examining this proportion, we are to use the mean distances, or the transverse semi-axes of the orbits (by prop. 15), and to neglect those little fractions, which, in defining the orbits, may have arisen from the insensible errors of observation, or may be ascribed to other causes which we shall afterwards explain. And thus we shall always find the said proportion to hold exactly; for the distances of Saturn, Jupiter, Mars, the Earth, Venus, and Mercury, from the sun, drawn from the observations of astronomers, are, according to the computation of *Kepler*, as the numbers 951000, 519650, 152350, 100000, 72400, 38806; by the computation of *Bullialdus*, as the numbers 954198, 522520, 152350, 100000, 72398, 38585; and from the periodic times they come out 953806, 520116, 152399, 100000, 72333, 38710. Their distances, according to *Kepler* and *Bullialdus*, scarcely differ by any sensible quantity, and where they differ most the distances drawn from the periodic times fall in between them.

That the circum-terrestrial force likewise decreases in the duplicate proportion of the distances, I infer thus.

The mean distance of the moon from the centre of the earth, is, in semi-diameters of the earth, according to *Ptolomy*, *Kepler* in his *Ephemerides*, *Bullialdus*, *Hevelius* and *Ricciolus*, 59; according to *Flamsted*, $59\frac{1}{3}$; according to *Tycho*, $56\frac{1}{2}$; to *Vendelin*, 60; to *Copernicus*, $60\frac{1}{3}$; to *Kircher*, $62\frac{1}{2}$.

But *Tycho*, and all that follow his tables of refraction, making the refractions of the sun and moon (altogether against the nature of light) to exceed those of the fixed stars, and that by about four or five minutes in the horizon, did thereby augment the horizontal parallax of the moon by about the like number of minutes; that is, by about the 12th or 15th part of the whole parallax. Correct this error, and the distance will become 60 or 61 semi-diameters of the earth, nearly agreeing with what others have determined.

Let us, then, assume the mean distance of the moon 60 semi-diameters of the earth, and its periodic time in respect of the fixed stars $27^{d.}\ 7^{h.}\ 43'$, as astronomers have determined it. And (by cor. 6, prop. 4) a body revolved in our air, near the surface of the earth, supposed at rest, by means of a centripetal force which should be to the same force at the distance of the moon in the reciprocal duplicate proportion of the distances from the centre of the earth, that is, as 3600 to 1, would (secluding the resistance of the air) complete a revolution in $1^{h.}\ 24'\ 27''$.

Suppose the circumference of the earth to be 123249600 *Paris* feet, as has been determined by the late mensuration of the *French*, then the same body, deprived of its circular motion, and falling by the impulse of the same centripetal force as before, would, in one second of time, describe $15\frac{1}{12}$ *Paris* feet.

This we infer by a calculus formed upon prop. 36, and it agrees with what we observe in all bodies about the earth. For by the experiments of pendulums, and a computation raised thereon, Mr. *Huygens* has demonstrated that bodies falling by all that centripetal force with which (of whatever nature it is) they are impelled near the surface of the earth, do, in one second of time, describe $15\frac{1}{12}$ *Paris* feet....

In such bodies as are found on our earth of very different sorts, I examine this analogy with great accuracy.

If the action of the circum-terrestrial force is proportional to the bodies to be moved, it will (by the second law of motion) move them with equal velocity in equal times, and will make all bodies let fall to descend through equal spaces in equal times, and all bodies hung by equal threads to vibrate in equal times. If the action of the force was greater, the times would be less; if that was less, these would be greater.

But it has long ago been observed by others, that (allowance being made for the small resistance of the air) all bodies descend through equal spaces in equal times; and, by the help of pendulums, that equality of times may be distinguished to great exactness.

I tried the thing in gold, silver, lead, glass, sand, common salt, wood, water, and wheat. I provided two equal wooden boxes. I filled the one with wood, and suspended an equal weight of gold (as exactly as I could) in the centre of oscillation of the other. The boxes, hung by equal threads of 11 feet, made a couple of pendulums perfectly equal in weight and figure, and equally exposed to the resistance of the air: and, placing the one by the other, I observed them to play together forwards and backwards for a long while, with equal vibrations. And therefore (by cor. 1 and 6, prop. 24, book 2) the quantity of matter in the gold was to the quantity of matter in the wood as the action of the motive force upon all the gold to the action of the same upon all the wood; that is, as the weight of one to the weight of the other.

And by these experiments, in bodies of the same weight, I could have discovered a difference of matter less than the thousandth part of the whole.

Since the action of the centripetal force upon the bodies attracted is, at equal distances, proportional to the quantities of matter in those bodies, reason requires that it should be also proportional to the quantity of matter in the body attracting.

For all action is mutual, and (by the third law of motion) makes the bodies mutually to approach one to the other, and therefore must be the same in both bodies. It is true that we may regard one body as attracting, another as attracted; but this distinction is more mathematical than natural. The attraction is really common of either to other, and therefore of the same kind in both.

And hence it is that the attractive force is found in both. The sun attracts Jupiter and the other planets; Jupiter attracts its satellites; and, for the same reason, the satellites act as well upon one another as upon Jupiter, and all the planets mutually one upon another....

Perhaps it may be objected, that, according to this philosophy all bodies should mutually attract one another, contrary to the evidence of experiments in terrestrial bodies; but I answer, that the experiments in terrestrial bodies come to no account; for the attraction of homogeneous spheres near their surfaces are (by prop. 72) as their diameters. Whence a sphere of one foot in diameter, and of a like nature to the earth, would attract a small body placed near its surface with a force 20000000 times less than the earth would do if placed near its surface; but so small a force could produce no sensible effect. If two such spheres were distant but by $\frac{1}{4}$ of an inch, they would not, even in spaces void of resistance, come together by the force of their mutual attraction in less than a month's time; and less spheres will come together at a rate yet slower, viz. in the proportion of their diameters. Nay, whole mountains will not be sufficient to produce any sensible effect. A mountain of an hemispherical figure, three miles high, and six broad, will not, by its attraction, draw the pendulum two minutes out of the true perpendicular; and it is only in the great bodies of the planets that these forces are to be perceived....

As the parts of the earth mutually attract one another, so do those of all the planets. If Jupiter and its satellites were brought together, and formed into one globe, without doubt they would continue mutually to attract one another as before. And, on the other hand, if the body of Jupiter was broke into more globes, to be sure, these would no less attract one another than they do the satellites now. From these attractions it is that the bodies of the earth and all the planets effect a spherical figure, and their parts cohere, and are not dispersed through the æther. But we have before proved that these forces arise from the universal nature of matter, and that, therefore, the force of any whole globe is made up of the several forces of all its parts. And from thence it follows (by cor. 3, prop. 74) that the force of every particle decreases in the duplicate proportion of the distance from that particle; and (by prop. 73 and 75) that the force of an entire globe, reckoning from the surface outwards, decreases in the duplicate, but, reckoning inwards, in the simple proportion of the distances from the centre, if the matter of the globe be uniform. And though the matter of the globe, reckoning from the centre towards the surface, is not uniform, yet the decrease in the duplicate proportion of the distance outwards would (by prop. 76) take place, provided that difformity is similar in places round about at equal distances from the centre. And two such globes will (by the same proposition) attract one the other with a force decreasing in the duplicate proportion of the distance between their centres.

Wherefore the absolute force of every globe is as the quantity of matter which the globe contains; but the motive force by which every globe is attracted towards another, and which, in terrestrial bodies, we commonly call their weight, is as the content under the quantities of matter in both globes applied to the square of the distance between their centres (by cor. 4, prop. 76), to which force the quantity of motion, by which each globe in a given time will be carried towards the other, is proportional. And the accelerative force, by which every globe according to its quantity of matter is attracted towards another, is as the quantity of matter in that other globe applied to the square of the distance between the centres of the two (by cor. 2, prop. 76); to which force, the velocity by which the attracted globe will, in a given time, be carried towards the other is proportional. And from these principles well understood, it will now be easy to determine the motions of the celestial bodies among themselves....

Thus I have given an account of the system of the planets. As to the fixed stars, the smallness of their annual parallax proves them to be removed to immense distances from the system of the planets: that this parallax is less than one minute is most certain; and from thence it follows that the distance of the fixed stars is above 360 times greater than the distance of Saturn from the sun. Such as reckon the earth one of the planets, and the sun one of the fixed stars, may remove the fixed stars to yet greater distances by the following arguments: from the annual motion of the earth there would happen an apparent transposition of the fixed stars, one in respect of another, almost equal to their double parallax; but the greater and nearer stars, in respect of the more remote, which are only seen by the telescope, have not hitherto been observed to have the least motion. If we should suppose that motion to be but less than 20″, the distance of the nearer fixed stars would exceed the mean distance of Saturn by above 2000 times....

The fixed stars being, therefore, at such vast distances from one another, can neither attract each other sensibly, nor be attracted by our sun. But the comets must unavoidably be acted on by the circum-solar force; for as the comets were placed by astronomers above the moon, because they were found to have no diurnal parallax, so their annual parallax is a convincing proof of their descending into the regions of the planets. For all the comets which move in a direct course, according to the order of the signs, about the end of their appearance become more than ordinarily slow, or retrograde, if the earth is between them and the sun; and more than ordinarily swift if the earth is approaching to a heliocentric opposition with them. Whereas, on the other hand, those which move against the order of the signs, towards the end of their appearance, appear swifter than they ought to be if the earth is between them and the sun; and slower, and perhaps retrograde, if the earth is in the other side of its orbit. This is occasioned by the motion of the earth in different situations. If the earth go the same way with the comet, with a swifter motion, the comet becomes retrograde; if with a slower motion, the comet becomes slower however; and if the earth move the contrary way, it becomes swifter; and by collecting the differences between the slower and swifter motions, and the sums of the more swift and retrograde motions, and comparing them with the situation and motion of the earth from whence they arise, I found, by means of this parallax, that the distances of the comets at the time they cease to be visible to the naked eye are always less than the distance of Saturn, and generally even less than the distance of Jupiter.

LAPLACE

THE complete exposition of the Newtonian theory, and its application to explain both the permanent stability and the periodic variations which are found in the solar system, are largely the work of two eminent Frenchmen—Lagrange and Laplace.

Laplace, especially, examined mathematically the effect of the attractions of the planets on each other, and deduced their observed perturbations from their mean orbits. He collected all known knowledge on the subject in his *Mécanique Céleste* and the more popular book, *Exposition du Système du Monde*. Passing to speculation, he framed the Nebular Hypothesis to explain the origin of the stupendous machine of the solar system, a hypothesis to some extent anticipated by Kant, and carried to much greater detail in more recent years.

Laplace, the son of a small farmer, was born in Normandy in 1749, and after adjusting his political opinions successfully to the changing fortunes of the time, was made a Marquis of the Restoration and died at Arcueil in 1827.

The SYSTEM OF THE WORLD

By M. Le Marquis de Laplace
Paris, 1796.

Translated from the French...by the Rev. Henry H. Harte, F.T.C.D., M.R.I.A.
Dublin, 1830.

From Book V. Summary of the History of Astronomy

Chap. VI. *Considerations on the system of the world, and on the future progress of astronomy*

The preceding summary of the history of Astronomy presents three distinct periods, which referring to the phenomena, to the laws which govern them, and to the forces on which these laws depend, point out the career of this science during its progress, and which consequently ought to be pursued in the cultivation of other sciences. The first period embraces the observations made by Astronomers antecedently to Copernicus, on the appearances of the celestial motions, and the hypotheses which were devised to explain those appearances, and to subject them to computation. In the second period, Copernicus deduced from these appearances, the motions of the Earth on its axis and about the Sun, and Kepler discovered the laws of the planetary motions. Finally in the third period, Newton, assuming the existence of these laws, established the principle of universal gravitation; and subsequent Geometers, by applying analysis to this principle, have derived from it all the observed phenomena, and the various inequalities in the motion of the planets, the satellites, and the comets. Astronomy thus becomes the solution of a great problem of mechanics, the constant arbitraries of which are the elements of the heavenly motions. It has all the certainty which can result from the immense number and variety of phenomena, which it rigorously explains, and from the simplicity of the principle which serves to explain them. Far from being apprehensive that the discovery of a new star will falsify this principle, we may be antecedently certain that its motion will be conformable to it; indeed this is what we ourselves have experienced with respect to Uranus and the four telescopic stars recently discovered, and every new comet which appears, furnishes us with an additional proof.

Such is unquestionably the constitution of the solar system. The immense globe of the Sun, the focus of these motions, revolves upon its axis in twenty-five days and a half. Its surface is covered with an ocean of luminous matter. Beyond it the planets, with their satellites, move, in orbits nearly circular, and in planes little inclined to the ecliptic. Innumerable comets, after having approached the Sun, recede to distances, which evince that his empire extends beyond the known limits of the planetary system. This luminary not only acts by its attraction upon all these globes, and compels them to move around him, but imparts to them both light and heat, his benign influence gives birth to the animals and plants which cover the surface of the Earth, and analogy induces us to believe, that he produces similar effects on the planets; for, it is not natural to suppose that matter, of which we see the fecundity develope itself in such various ways, should be sterile upon such a planet as Jupiter, which, like the Earth, has its days, its nights, and its years, and on which observation discovers changes that indicate very active forces. Man, formed for the temperature which he enjoys upon the Earth, could not, according to all appearance, live upon the other planets; but ought there not to be a diversity of organization suited to the various temperatures of the globes of this universe? If the difference of elements and climates alone causes such variety in the productions of the Earth, how infinitely diversified must be the productions of the planets and their satellites? The most active imagination cannot form any just idea of them, but still their existence is, at least, extremely probable.

However arbitrary the elements of the system of the planets may be, there exists between them some very remarkable relations, which may throw light on their origin. Considering it with attention, we are astonished to see all the planets move round the Sun from west to east, and nearly in the same plane, all the satellites moving round their respective planets in the same direction, and nearly in the same plane with the planets. Lastly, the Sun, the planets, and those satellites in which a motion of rotation have been observed, turn on their own axes, in the same direction, and nearly in the same plane as their motion of projection.

The satellites exhibit in this respect a remarkable peculiarity. Their motion of rotation is exactly equal to their motion of revolution; so that they always present the same hemisphere to their primary. At least, this has been observed for the Moon, for the four satellites of Jupiter, and for the last satellite of Saturn, the only satellites whose rotation has hitherto been recognized.

Phenomena so extraordinary, are not the effect of irregular causes. By subjecting their probability to computation, it is found that there is more than two thousand to one against the hypothesis that they are the effect of chance, which is a probability much greater than that on which most of the events of history, respecting which there does not exist a doubt, depends. We ought therefore to be assured with the same confidence, that a primitive cause has directed the planetary motions.

Another phenomenon of the solar system equally remarkable, is the small excentricity of the orbits of the planets and their satellites, while those of comets are very much extended. The orbits of this system present no intermediate shades between a great and small excentricity. We are here again compelled to acknowledge the effect of a regular cause; chance alone could not have given a form nearly circular to the orbits of all the planets. It is therefore necessary that the cause which determined the motions of these bodies, rendered them also nearly circular. This cause then must also have influenced the great excentricity of the orbits of comets, and their motion in every direction; for, considering the orbits of retrograde comets, as being inclined more than one hundred degrees to the ecliptic, we find that the mean inclination of the orbits of all the observed comets, approaches near to one hundred degrees, which would be the case if the bodies had been projected at random.

What is this primitive cause? In the concluding note of this work I will suggest an hypothesis which appears to me to result with a great degree of probability, from the preceding phenomena, which however I present with that diffidence, which ought always to attach to whatever is not the result of observation and computation.

Whatever be the true cause, it is certain that the elements of the planetary system are so arranged as to enjoy the greatest possible stability, unless it is deranged by the intervention of foreign causes. From the sole circumstance that the motions of the planets and satellites are performed in orbits nearly circular, in the same direction, and in planes which are inconsiderably inclined to each other, the system will always oscillate about a mean state, from which it will deviate but by very small quantities. The mean motions of rotation and of revolution of these different bodies are uniform, and their mean distances from the foci of the principal forces which actuate them are constant; all the secular inequalities are periodic....

From: NOTE VII and last

Let us consider whether we can assign the true cause. Whatever may be its nature, since it has produced or influenced the direction of the planetary motions, it must have embraced them all within the sphere of its action and, considering the immense distance which intervenes between them, nothing could have effected this but a fluid of almost indefinite extent. In order to have impressed on them all a motion, circular and in the same direction about the Sun, this fluid must environ this star, like an atmosphere. From a consideration of the planetary motions, we are therefore brought to the conclusion, that in consequence of an excessive heat, the solar atmosphere originally extended beyond the orbits of all the planets, and that it has successively contracted itself within its present limits.

In the primitive state in which we have supposed the Sun to be, it resembles those substances which are termed nebulæ, which, when seen through telescopes, appear to be composed of a nucleus, more or less brilliant, surrounded by a nebulosity, which, by condensing on its surface, transforms it into a star. If all the stars are conceived to be similarly formed, we can suppose their anterior state of nebulosity to be preceded by other states, in which the nebulous matter was more or less diffuse, the nucleus being at the same time more or less brilliant. By going back in this manner, we shall arrive at a state of nebulosity so diffuse, that its existence can with difficulty be conceived.

For a considerable time back, the particular arrangement of

some stars visible to the naked eye, has engaged the attention of philosophers. Mitchel remarked long since how extremely improbable it was that the stars composing the constellation called the Pleiades, for example, should be confined within the narrow space which contains them, by the sole chance of hazard; from which he inferred that this group of stars, and the similar groups which the heavens present to us, are the effects of a primitive cause, or of a primitive law of nature. These groups are a general result of the condensation of nebulæ of several nuclei; for it is evident that the nebulous matter being perpetually attracted by these different nuclei, ought at length to form a group of stars, like to that of the Pleiades. The condensation of nebulæ consisting of two nuclei, will in like manner form stars very near to each other, revolving the one about the other like to the double stars, whose respective motions have been already recognized.

But in what manner has the solar atmosphere determined the motions of rotation and revolution of the planets and satellites? If these bodies had penetrated deeply into this atmosphere, its resistance would cause them to fall on the Sun. We may therefore suppose that the planets were formed at its successive limits, by the condensation of zones of vapours, which it must, while it was cooling, have abandoned in the plane of its equator....

If all the particles of a ring of vapours continued to condense without separating, they would at length constitute a solid or a liquid ring. But the regularity which this formation requires in all parts of the ring, and in their cooling, ought to make this phenomenon very rare. Thus the solar system presents but one example of it; that of the rings of Saturn. Almost always each ring of vapours ought to be divided into several masses, which, being moved with velocities which differ little from each other, should continue to revolve at the same distance about the Sun. These masses should assume a spheroidical form, with a rotatory motion in the direction of that of their revolution, because their inferior particles have a less real velocity than the superior; they have therefore constituted so many planets in a state of vapour. But if one of them was sufficiently powerful, to unite successively by its attraction, all the others about its centre, the ring of vapours would be changed into one sole spheroidical mass, circulating about the Sun, with a motion of rotation in the same direction with that of revolution. This last case has been the most common; however, the solar system presents to us the first case, in the four small planets which revolve between Mars and Jupiter, at least unless we suppose with Olbers, that they originally formed one planet only, which was divided by an explosion into several parts, and actuated by different velocities. Now if we trace the changes which a farther cooling ought to produce in the planets formed of vapours, and of which we have suggested the formation, we shall see to arise in the centre of each of them, a nucleus increasing continually, by the condensation of the atmosphere which environs it. In this state, the planet resembles the Sun in the nebulous state, in which we have first supposed it to be; the cooling should therefore produce at the different limits of its atmosphere, phenomena similar to those which have been described, namely, rings and satellites circulating about its centre in the direction of its motion of rotation, and revolving in the same direction on their axes. The regular distribution of the mass of rings of Saturn about its centre and in the plane of its equator, results naturally from this hypothesis, and, without it, is inexplicable. Those rings appear to me to be existing proofs of the primitive extension of the atmosphere of Saturn, and of its successive condensations. Thus the singular phenomena of the small eccentricities of the orbits of the planets and satellites, of the small inclination of these orbits to the solar equator, and of the identity in the direction of the motions of rotation and revolution of all those bodies with that of the rotation of the Sun, follow from the hypothesis which has been suggested, and render it extremely probable. If the solar system was formed with perfect regularity, the orbits of the bodies which compose it would be circles, of which the planes, as well as those of the various equators and rings, would coincide with the plane of the solar equator. But we may suppose that the innumerable varieties which must necessarily exist in the temperature and density of different parts of these great masses, ought to produce the eccentricities of their orbits, and the deviations of their motions, from the plane of this equator.

In the preceding hypothesis, the comets do not belong to the solar system. If they be considered, as we have done, as small nebulæ, wandering from one solar system to another, and formed by the condensation of the nebulous matter, which is diffused so profusely throughout the universe, we may conceive that when they arrive in that part of space where the attraction of the Sun predominates, it should force them to describe elliptic or hyperbolic orbits. But as their velocities are equally possible in every direction, they must move indifferently in all directions, and at every possible inclination to the ecliptic; which is conformable to observation. Thus the condensation of the nebulous matter, which explains the motions of rotation and revolution of the planets and satellites in the same direction, and in orbits very little inclined to each other, likewise explains why the motions of the comets deviate from this general law.

FOUCAULT, STOKES, BUNSEN AND KIRCHHOFF

THE old distinction in kind between the celestial and the terrestrial spheres was broken down by Galileo and Newton. The mechanical laws which describe the movement of falling bodies on the surface of the earth were found to hold good in the farthest limits of the solar system. Experiments and calculations made in a laboratory, could elucidate the majestic motion of planets in their orbits.

To complete the demonstration of identity, however, it is necessary to show similarity of structure and composition as well as of motion. Are the chemical elements, from which all terrestrial things are made, the basis also of sun, planets and stars? It may well have seemed a question hopeless to ask. Yet a solution was found along a road opened up by Newton himself. He it was who discovered the coloured band of light produced by passing the sun's rays through a prism. This solar spectrum is crossed by a number of dark lines, while the light from colourless flames tinged with metals or salts shows bright lines on a dark field. These bright and dark lines were found to coincide—the spectrum of sodium for example giving two bright lines coincident with two dark lines in the sun's light.

The explanation of these facts was given independently by M. Léon Foucault, by Professor Gustav Robert Kirchhoff of Heidelberg and by Sir George Gabriel Stokes, who held Newton's old professorial chair at Cambridge. They showed that this spectrum analysis gave a proof that terrestrial chemical elements existed in the sun and the luminous stars. The heavenly bodies were as much and no more incorruptible, divine, than the familiar though mysterious matter of this our globe. The boundless depths of space lay open yet further to our investigation.

The principle underlying this method of spectrum analysis is that of resonance. If a note be sung near a piano, the particular wires tuned to that note will be set in vibration. That is to say, they will absorb vibrational energy of the same period as they themselves emit. If a complex sound passed through a grove of piano wires, and were analysed on the further side, it would be found to be wanting in those particular notes which the wires, if more vigorously excited, would send forth. And so the light of the sun, passing through his cooler envelope, is deprived of those rays the constituents of that envelope would emit, and the coloured band of the solar spectrum is crossed by dark lines indicating the elements present in the sun's atmosphere. This explanation, modestly given in his lectures by Stokes, made the name of other men.

EXTRACTS RELATING TO THE EARLY HISTORY OF SPECTRUM ANALYSIS

On the Simultaneous Emission and Absorption of Rays of the same definite Refrangibility; being a translation of a portion of a paper by M. Léon Foucault, and of a paper by Professor Kirchhoff.

(From the *Philosophical Magazine*, xix. (1860), pp. 196–7.)

Gentlemen,

Some years ago M. Foucault mentioned to me in conversation a most remarkable phænomenon which he had observed in the course of some researches on the voltaic arc, but which, though published in *L'Institut*, does not seem to have attracted the attention which it deserves. Having recently received from Prof. Kirchhoff a copy of a very important communication to the Academy of Sciences at Berlin, I take the liberty of sending you translations of the two, which I doubt not will prove highly interesting to many of your readers.

I am, Gentlemen, Yours sincerely,

G. G. STOKES.

M. Foucault's discovery is mentioned in the course of a paper published in *L'Institut* of Feb. 7, 1849, having been brought forward at a meeting of the Philomathic Society on the 20th of January preceding. In describing the result of a prismatic analysis of the voltaic arc formed between charcoal poles, M. Foucault writes as follows (p. 45):—

"Its spectrum is marked, as is known, in its whole extent by a multitude of irregularly grouped luminous lines: but among these may be remarked a double line situated at the boundary of the yellow and orange. As this double line recalled by its form and situation the line *D* of the solar spectrum, I wished to try if it corresponded to it; and in default of instruments for measuring the angles, I had recourse to a particular process.

"I caused an image of the sun, formed by a converging lens, to fall on the arc itself, which allowed me to observe at the same time the electric and the solar spectrum superposed; I convinced myself in this way that the double bright line of the arc coincides exactly with the double dark line of the solar spectrum.

"This process of investigation furnished me matter for some unexpected observations. It proved to me in the first instance the extreme transparency of the arc, which occasions only a faint shadow in the solar light. It showed me that this arc, placed in the path of a beam of solar light, absorbs the rays *D*, so that the above mentioned line *D* of the solar light is considerably strengthened when the two spectra are exactly superposed. When, on the contrary, they jut out one beyond the other, the line *D* appears darker than usual in the solar light, and stands out bright in the electric spectrum, which allows one easily to judge of their perfect coincidence. Thus the arc presents us with a medium which emits the rays *D* on its own account, and which at the same time absorbs them when they come from another quarter.

"To make the experiment in a manner still more decisive, I projected on the arc the reflected image of one of the charcoal points, which, like all solid bodies in ignition, gives no lines; and under these circumstances the line *D* appeared to me as in the solar spectrum."

Prof. Kirchhoff's communication "On Fraunhofer's Lines," dated Heidelberg, 20th of October, 1859, was brought before the Berlin Academy on the 27th of that month, and is printed in the *Monatsbericht*, p. 662.

"On the occasion of an examination of the spectra of coloured flames not yet published, conducted by Bunsen and myself in common, by which it has become possible for us to recognise the qualitative composition of complicated mixtures from the appearance of the spectrum of their blowpipe-flame, I made some observations which disclose an unexpected explanation of the origin of Fraunhofer's lines, and authorise conclusions therefrom respecting the material constitution of the atmosphere of the sun, and perhaps also of the brighter fixed stars.

"Fraunhofer had remarked that in the spectrum of the flame of a candle there appear two bright lines, which coincide with the two dark lines *D* of the solar spectrum. The same bright lines are obtained of greater intensity from a flame in which some common salt is put. I formed a solar spectrum by projection, and allowed the solar rays concerned, before they fell on the slit, to pass through a powerful salt-flame. If the sunlight were sufficiently reduced, there appeared in place of the two dark lines *D* two bright lines: if, on the other hand, its intensity surpassed a certain limit, the two dark lines *D* showed themselves in much greater distinctness than without the employment of the salt-flame.

"The spectrum of the Drummond light contains, as a general rule, the two bright lines of sodium, if the luminous spot of the cylinder of lime has not long been exposed to the white heat; if the cylinder remains unmoved these lines become weaker, and finally vanish altogether. If they have vanished, or only faintly appear, an alcohol flame into which salt has been put, and which is placed between the cylinder of lime and the slit, causes two dark lines of remarkable sharpness and fineness, which in that respect agree with the lines *D* of the solar spectrum, to show themselves in their stead. Thus the lines *D* of the solar spectrum are artificially evoked in a spectrum in which naturally they are not present.

"If chloride of lithium is brought into the flame of Bunsen's gas-lamp, the spectrum of the flame shows a very bright sharply defined line, which lies midway between Fraunhofer's lines *B* and *C*. If, now, solar rays of moderate intensity are allowed to fall through the flame on the slit, the line at the place pointed out is seen bright on a darker ground; but with greater strength of sunlight there appears in its place a dark line, which has quite the same character as Fraunhofer's lines. If the flame be taken away, the line disappears, as far as I have been able to see, completely.

"I conclude from these observations, that coloured flames, in the spectra of which bright sharp lines present themselves, so weaken rays of the colour of these lines, when such rays pass through the flames, that in place of the bright lines dark ones appear as soon as there is brought behind the flame a source of light of sufficient intensity, in the spectrum of which these lines are otherwise wanting. I conclude further, that the dark lines of the solar spectrum which are not evoked by the atmosphere of the earth, exist in consequence of the presence, in the incandescent atmosphere of the sun, of those substances which in the spectrum of a flame produce bright lines at the same place. We may assume that the bright lines agreeing with *D* in the spectrum of a flame always arise from sodium contained in it; the dark line *D* in the solar spectrum allows us, therefore, to conclude that there exists sodium in the sun's atmosphere. Brewster has found bright lines in the spectrum of the flame of saltpeter at the place of Fraunhofer's lines *A*, *a*, *B*; these lines point to the existence of potassium in the sun's atmosphere. From my observation, according to which no dark line in the solar spectrum answers to the red line of lithium, it would follow with probability that in the atmosphere of the sun lithium is either absent, or is present in comparatively small quantity.

"The examination of the spectra of coloured flames has accordingly acquired a new and high interest; I will carry it out in conjunction with Bunsen as far as our means allow. In connexion therewith we will investigate the weakening of rays of light in flames that has been established by my observations. In the course of the experiments which have at present been instituted by us in this direction, a fact has already shown itself which seems to us to be of great importance. The Drummond light requires, in order that the lines *D* should come out in it dark, a salt-flame of lower temperature. The flame of alcohol containing water is fitted for this, but the flame of Bunsen's gas-lamp is not. With the latter the smallest mixture of common salt, as soon as it makes itself generally perceptible, causes the bright lines of sodium to show themselves. We reserve to ourselves to develope the consequences which may be connected with this fact."

EINSTEIN, MINKOWSKI, EDDINGTON

RELATIVITY

Light has the properties of wave motion, and hence there has been imagined a medium or aether to carry it. Even if we suppose the solar system at rest in the aether of space, the earth in its daily and annual path must move through the aether. Astronomical observation indicates that it does not drag aether with it, and therefore we must imagine the aether to stream through a laboratory as wind through a grove of trees. We should expect that light reflected back by a mirror would travel with different speeds in different directions, according as its path was up and down or across and across the aether stream. But an experiment made in America by Michelson and Morley in 1887 gave the surprising result that no such difference could be detected. However measured, and in whatever direction, the speed of light always appears to be 186,000 miles a second.

If we try to explain this constancy in terms of our old conceptions of the world, we are driven to suppose compensating changes in the scales and clocks with which the measurements are made. The electrical theory of matter gives reasons for this assumption, and it probably represents one aspect of the truth. But in 1905 Einstein founded the modern theory of relativity by pointing out that absolute space and time are figments of the imagination, and that real or observed space and time as measured depend on the motion of the observer and his instruments.

In 1908 Minkowski showed that if we cease to think of space and time as separate entities, and bring the length, breadth and thickness of space into relation with time, we get a four-dimensional continuum which is the same for all observers. In the old concept of the world, the distance between two points was an absolute quantity however measured. In the new four-dimensional world that distance will depend on the observer, but there is an absolute quantity, an extension in space and time combined, or "interval," which is the same for all observers, though they may divide it into space and time in different manners and with different results. The natural connection between time and space is given by the fact that light travels 186,000 miles in a second. Hence, in the four-dimensional world, one second is equivalent to 186,000 miles, and the discrepancies in our old conceptions of the world only become apparent when great velocities are involved.

There is still a difference—one of sign—between time and the three space dimensions. The mathematical consequence is that the space-time continuum of our new world does not give the geometry made familiar by Euclid. It is subject to something analogous to curvature, and, if a groove or pleat is started, it can only run in one direction through the continuum. Thus there are natural paths in this space-time which, on the theory, are the tracks of freely moving particles.

In 1911 Einstein showed the bearing of these conceptions on gravitation. The extra weight we feel as a lift starts, or the centrifugal force of a whirling stone, are forces of the same nature as that of gravity. It is impossible to draw a distinction in kind between their effects—they are equivalent. Thus the effects of gravity may be imitated exactly by an acceleration or a change of path. A curvature of space, taking bodies out of their free paths, may therefore be the cause of the phenomenon we call gravitation.

Gravity affects not only material bodies, but light as well. If light passes near a heavy body such as the Sun, it is deflected from its straight path, and the deflection should be twice as great on Einstein's theory as on Newton's. Professor Eddington, who has done much to extend and expound Einstein's theory, working with other observers, carried out the delicate astronomical measurements needed to test this crucial point, and found that the facts conformed to Einstein's view and not to Newton's.

SPACE, TIME AND GRAVITATION

An Outline of the General Relativity Theory

By A. S. Eddington, M.A., M.Sc., F.R.S.

Plumian Professor of Astronomy and
Experimental Philosophy, Cambridge.

Cambridge: 1921

Chap. VI. *The New Law of Gravitation and the Old Law*

I don't know what I may seem to the world, but, as to myself, I seem to have been only as a boy playing on the sea-shore, and diverting myself in now and then finding a smoother pebble or a prettier shell than ordinary, whilst the great ocean of truth lay all undiscovered before me.

Sir Isaac Newton.

Was there any reason to feel dissatisfied with Newton's law of gravitation?

Observationally it had been subjected to the most stringent tests, and had come to be regarded as the perfect model of an exact law of nature. The cases, where a possible failure could be alleged, were almost insignificant. There are certain unexplained irregularities in the moon's motion; but astronomers generally looked—and must still look—in other directions for the cause of these discrepancies. One failure only had led to a serious questioning of the law; this was the discordance of motion of the perihelion of Mercury. How small was this discrepancy may be judged from the fact that, to meet it, it was proposed to amend *square* of the distance to the 2·00000016 power of the distance. Further it seemed possible, though unlikely, that the matter causing the zodiacal light might be of sufficient mass to be responsible for this effect.

The most serious objection against the Newtonian law as an exact law was that it had become ambiguous. The law refers to the product of the masses of the two bodies; but the mass depends on the velocity—a fact unknown in Newton's day. Are we to take the variable mass, or the mass reduced to rest? Perhaps a learned judge, interpreting Newton's statement like a last will and testament, could give a decision; but that is scarcely the way to settle an important point in scientific theory.

Further *distance*, also referred to in the law, is something relative to an observer. Are we to take the observer travelling with the sun or with the other body concerned, or at rest in the aether or in some gravitational medium?...

It is often urged that Newton's law of gravitation is much simpler than Einstein's new law. That depends on the point of view; and from the point of view of the four-dimensional world Newton's law is far more complicated. Moreover, it will be seen that if the ambiguities are to be cleared up, the statement of Newton's law must be greatly expanded.

Some attempts have been made to expand Newton's law on the basis of the restricted principle of relativity alone. This was insufficient to determine a definite amendment. Using the principle of equivalence, or relativity of force, we have arrived at a definite law proposed in the last chapter. Probably the question has arisen in the reader's mind, why should it be called the law of gravitation? It may be plausible as a law of nature; but what has the degree of curvature of space-time to do with attractive forces, whether real or apparent?

A race of flat-fish once lived in an ocean in which there were only two dimensions. It was noticed that in general fishes swam in straight lines, unless there was something obviously interfering with their free courses. This seemed a very natural behaviour. But there was a certain region where all the fish seemed to be bewitched; some passed through the region but changed the direction of their swim, others swam round and round indefinitely. One fish invented a theory of vortices, and said that there were whirlpools in that region which carried everything round in curves. By-and-by a far better theory was proposed; it was said that the fishes were all attracted towards a particularly large fish—a sun-fish—which was lying asleep in the middle of the region; and that was what caused the deviation of their paths. The theory might not have sounded particularly plausible at first; but it was confirmed by marvellous exactitude by all kinds of experimental tests. All fish were found to possess this attractive power in proportion to their sizes; the law of attraction was extremely simple, and yet it was found to explain all the motions with an accuracy never approached before in any scientific investigations. Some fish grumbled that they did not see how there could be such an influence at a distance; but it was generally agreed that the influence was communicated through the ocean and might be better understood when more was known about the nature of water. Accordingly, nearly every fish who wanted to explain the attraction started by proposing some kind of mechanism for transmitting it through the water.

But there was one fish who thought of quite another plan. He was impressed by the fact that whether the fish were big or little they always took the same course, although it would naturally take a bigger force to deflect the bigger fish. He therefore concentrated on the courses rather than on the forces. And then he arrived at a striking explanation of the whole thing. There was a mound in the world round about where the sun-fish lay. Flat-fish could not appreciate it directly because they were two-dimensional; but whenever a fish went swimming over the slopes of the mound, although he did his best to swim straight on, he got turned round a bit. (If a traveller goes over the left slope of a mountain, he must consciously keep bearing away to the left if he wishes to keep to his original direction relative to the points of the compass.) This was the secret of the mysterious attraction, or bending of the paths, which was experienced in the region.

The parable is not perfect, because it refers to a hummock in space alone, whereas we have to deal with hummocks in space-time. But it illustrates how a curvature of the world we live in may give an illusion of attractive force, and indeed can only be discovered through some such effect. How this works out in detail must now be considered.

In the form $G_{\mu\nu} = 0$, Einstein's law expresses conditions to be satisfied in a gravitational field produced by any arbitrary distribution of attracting matter. An analogous form of Newton's law was given by Laplace in his celebrated expression $\nabla^2 V = 0$. A more illuminating form of the law is obtained if, instead of putting the question what kinds of space-time can exist under the most general conditions in an empty region, we ask what kind of space-time exists in the region round a single attracting particle? We separate out the effect of a single particle, just as Newton did....

We need only consider space of two dimensions—sufficient for the so-called plane orbit of a planet—time being added as the third dimension. The remaining dimension of space can always be added, if desired, by conditions of symmetry. The result of long algebraic calculations is that, round a particle

$$ds^2 = -\frac{1}{\gamma}dr^2 - r^2 d\theta^2 + \gamma dt^2 \quad \text{.........}(6)$$

where $$\gamma = 1 - \frac{2m}{r}.$$

The quantity m is the gravitational mass of the particle—but we are not supposed to know that at present. r and θ are polar coordinates, or rather they are the nearest thing to polar coordinates that can be found in space which is not truly flat.

The fact is that this expression for ds^2 is found in the first place simply as a particular solution of Einstein's equations of the gravitational field; it is a variety of hummock (apparently the simplest variety) which is not curved beyond the first degree. There *could* be such a state of the world under suitable circumstances. To find out what those circumstances are, we have to trace some of the consequences, find out how any particle moves when ds^2 is of this form, and then examine whether we know of any case in which these circumstances are found observationally. It is only after having ascertained that this form of ds^2 does correspond to the leading observed effects attributable to a particle of mass m at the origin that we have the right to identify this particular solution with the one we hoped to find.

It will be a sufficient illustration of this procedure, if we indicate how the position of the matter causing this particular solution is located. Wherever the formula (6) holds good there can be no matter, because the law which applies to empty space is satisfied. But if we try to approach the origin ($r = 0$), a curious thing happens. Suppose we take a measuring-rod, and, laying it radially, start marking off equal lengths with it along a radius, gradually approaching the origin. Keeping the time t constant, and $d\theta$ being zero for radial measurements, the formula (6) reduces to

$$ds^2 = -\frac{1}{\gamma}dr^2$$

or $$dr^2 = -\gamma ds_2.$$

We start with r large. By-and-by we approach the point where $r = 2m$. But here, from its definition, γ is equal to 0. So that, however large the measured interval ds may be, $dr = 0$. We can go on shifting the measuring-rod through its own length time after time, but dr is zero; that is to say, we do not reduce r. There is a magic circle which no measurement can bring us inside. It is not unnatural that we should picture something obstructing our closer approach, and say that a particle of matter is filling up the interior.

The fact is that so long as we keep to space-time curved only in the first degree, we can never round off the summit of the hummock. It must end in an infinite chimney. In place of the chimney, however, we round it off with a small region of greater curvature. This region cannot be empty because the law applying to empty space does not hold. We describe it therefore as containing matter—a procedure which practically amounts to a definition of matter. Those familiar with hydrodynamics may be reminded of the problem of the irrotational rotation of a fluid; the conditions cannot be satisfied at the origin, and it is necessary to cut out a region which is filled by a vortex filament.

A word must also be said as to the co-ordinates r and t used in (6). They correspond to our ordinary notion of radial distance and time—as well as any variables in a non-Euclidean world can correspond to words which, as ordinarily used, presuppose a Euclidean world. We shall thus call r and t, distance and time. But to give names to coordinates does not give more information—and in this case gives considerably less information—than is already contained in the formula for ds^2. If any question arises as to the exact significance of r and t it must always be settled by reference to equation (6).

The want of flatness in the gravitational field is indicated by the deviation of the coefficient γ from unity. If the mass $m = 0$, $\gamma = 1$, and space-time is perfectly flat. Even in the most intense gravitational field known, the deviation is extremely small. For the sun, the quantity m, called the gravitational mass, is only 1·47 kilometres, for the earth it is 5 millimetres. In any practical problem the ratio $2m/r$ must be exceedingly small. Yet it is on the small corresponding difference in r that the whole of the phenomena of gravitation depend....

The mathematical reader should find no difficulty in proving that for a particle with small velocity the acceleration towards the sun is approximately m/r^2, agreeing with the Newtonian law....

The result that the expression found for the geometry of the gravitational field of a particle leads to Newton's law of attraction is of great importance. It shows that the law $G_{\mu\nu} = 0$, proposed on theoretical grounds, agrees with observation at least approximately. It is no drawback that the Newtonian law only applies when the speed is small; all planetary speeds are small compared with the velocity of light, and the considerations mentioned at the beginning of this chapter suggest that some modification may be needed for speeds comparable with that of light.

Another important point to notice is that the attraction of gravitation is simply a geometrical deformation of the straight tracks. It makes no difference what body or influence is pursuing the track, the deformation is a general discrepancy between the "mental picture" and the "true map" of the portion of space-time considered. Hence light is subject to the same disturbance of path as matter. This is involved in the Principle of Equivalence; otherwise we could distinguish between the acceleration of a lift and a true increase of gravitation by optical experiments; in that case the observer for whom light-rays appear to take straight tracks might be described as absolutely unaccelerated and there could be no relativity theory. Physicists in general have been prepared to admit the likelihood of an influence of gravitation on light similar to that exerted on matter; and the problem whether or no light has "weight" has often been considered.

The appearance of γ as the coefficient of dt^2 is responsible for the main features of Newtonian gravitation; the appearance of $1/\gamma$ as the coefficient of dr^2 is responsible for the principal deviations of the new law from the old. This classification seems to be correct; but the Newtonian law is ambiguous and it is difficult to say exactly what are to be regarded as discrepancies from it. Leaving aside now the time-term as sufficiently discussed, we consider the space-terms alone[1]

$$ds^2 = \frac{1}{\gamma} dr^2 + r^2 d\theta^2.$$

[1] We change the sign of ds^2, so that ds, when real, means measured space instead of measured time.

The expression shows that space considered alone is non-Euclidean in the neighbourhood of an attracting particle. This is something entirely outside the scope of the old law of gravitation. Time can only be explored by something moving, whether a free particle or the parts of a clock, so that the non-Euclidean character of space-time can be covered up by introducing a field of force, suitably modifying the motion, as a convenient fiction. But space can be explored by static methods; and theoretically its non-Euclidean character could be ascertained by sufficiently precise measures with rigid scales.

If we lay our measuring scale transversely and proceed to measure the circumference of a circle of nominal radius r, we see from the formula that the measured length ds is equal to $dr\theta$, so that, when we have gone right round the circle, θ has increased by 2π and the measured circumference is $2\pi r$. But when we lay the scale radially the measured length ds is equal to $dr/\sqrt{\gamma}$, which is always greater than dr. Thus, in measuring a diameter, we obtain a result greater than $2r$, each portion being greater than the corresponding change of r.

Thus if we draw a circle, placing a massive particle near the centre so as to produce a gravitational field, and measure with a rigid scale the circumference and the diameter, the ratio of the measured circumference to the measured diameter will not be the famous number

$$\pi = 3{\cdot}14159265358979323846264338327 9\ldots$$

but a little smaller. Or if we inscribe a regular hexagon in this circle its sides will not be exactly equal to the radius of the circle. Placing the particle near, instead of at, the centre, avoids measuring the diameter *through* the particle, and so makes the experiment a practical one. But though practical, it is not practicable to determine the non-Euclidean character of space in this way. Sufficient refinement of measures is not attainable. If the mass of a ton were placed inside a circle of five yards radius, the defect in the value of π would only appear in the twenty-fourth or twenty-fifth place of decimals.

It is of value to put the result in this way, because it shows that the relativist is not talking metaphysics when he says that space in the gravitational field is non-Euclidean. His statement has a plain physical meaning, which we may some day learn how to test experimentally. Meanwhile we can test it by indirect methods....

[A body passing near a massive particle has its path bent owing to the non-Euclidean character of space.]

This bending of the path is additional to that due to the Newtonian force of gravitation which depends on the second appearance of γ in the formula. As already explained it is in general a far smaller effect and will appear only as a minute correction to Newton's law. The only case where the two rise to equal importance is when the track is that of a light wave, or of a particle moving with a speed approaching that of light; for then dr^2 rises to the same order of magnitude as dt^2.

To sum up, a ray of light passing near a heavy particle will be bent, firstly, owing to the non-Euclidean character of the combination of time with space. This bending is equivalent to that due to Newtonian gravitation, and may be calculated in the ordinary way on the assumption that light has weight like a material body. Secondly, it will be bent owing to the non-Euclidean character of space alone, and this curvature is additional to that predicted by Newton's law. If then we can observe the amount of curvature of a ray of light, we can make a crucial test of whether Einstein's or Newton's theory is obeyed....

It is not difficult to show that the total deflection of a ray of light passing at a distance r from the centre of the sun is (in circular measure) $\frac{4m}{r}$, whereas the deflection of the same ray calculated on the Newtonian theory would be $\frac{2m}{r}$. For a ray grazing the surface of the sun the numerical value of this deflection is

$1''{\cdot}75$ (Einstein's theory),
$0''{\cdot}87$ (Newtonian theory)....

The bending affects stars seen near the sun, and accordingly the only chance of making the observation is during a total eclipse when the moon cuts off the dazzling light. Even then there is a great deal of light from the sun's corona which stretches far above the disc. It is thus necessary to have rather bright stars near the sun, which will not be lost in the glare of the corona. Further the displacements of these stars can only be measured relatively to other stars, preferably more distant from the sun and less displaced; we need therefore a reasonable number of outer bright stars to serve as reference points.

In a superstitious age a natural philosopher wishing to perform an important experiment would consult an astrologer to ascertain an auspicious moment for the trial. With better reason, an astronomer to-day consulting the stars would announce that the most favourable day of the year for weighing light is May 29. The reason is that the sun in its annual journey round the ecliptic goes through fields of stars of varying richness, but on May 29 it is in the midst of a quite exceptional patch of bright stars—part of the Hyades—by far the best star-field encountered. Now if this problem had been put forward at some other period of history, it might have been necessary to wait some thousands of years for a total eclipse of the sun to happen on the lucky date. But by strange good fortune an eclipse did happen on May 29, 1919. Owing to the curious sequence of eclipses a similar opportunity will recur in 1938; we are in the midst of the most favourable cycle. It is not suggested that it is impossible to make the test at other eclipses, but the work will necessarily be more difficult.

Attention was called to this remarkable opportunity by the Astronomer Royal in March, 1917; and preparations were begun by a Committee of the Royal Society and Royal Astronomical Society for making the observations. Two expeditions were sent to different places on the line of totality to minimise the risk of failure by bad weather. Dr A. C. D. Crommelin and Mr C. Davidson went to Sobral in North Brazil; Mr E. T. Cottingham and the writer went to the Isle of Principe in the Gulf of Guinea, West Africa....

It will be remembered that Einstein's theory predicts a deflection of $1''\cdot 74$ at the edge of the sun*, the amount falling off inversely as the distance from the sun's centre. The simple Newtonian deflection is half this, $0''\cdot 87$. The final results (reduced to the edge of the sun) obtained at Sobral and Principe with their "probable accidental errors" were

Sobral	$1''\cdot 98 \pm 0''\cdot 12$
Principe	$1''\cdot 61 \pm 0''\cdot 30$.

* The predicted deflection of light from infinity to infinity is just over $1'\cdot 745$; from infinity to the earth it is just under.

It is usual to allow a margin of safety of about twice the probable error on either side of the mean. The evidence of the Principe plates is thus just about sufficient to rule out the possibility of the "half-deflection," and the Sobral plates exclude it with practical certainty. The value of the material found at Principe cannot be put higher than about one-sixth of that at Sobral; but it certainly makes it less easy to bring criticism against this confirmation of Einstein's theory seeing that it was obtained independently with two different instruments at different places and with different kinds of checks.

The best check on the results obtained with the 4-inch lens at Sobral is the striking internal accordance of the measures for different stars. The theoretical deflection should vary inversely as the distance from the sun's centre; hence, if we plot the mean radial displacement found for each star separately against the inverse distance, the points should lie on a straight line. This is shown in Fig. 17 where the broken line shows the theoretical prediction of Einstein, the deviations being within the accidental errors of the determinations. A line of half the slope representing the half-deflection would clearly be inadmissible....

We have seen that the swift-moving light-waves possess great advantages as a means of exploring the non-Euclidean property of space. But there is an old fable about the hare and the tortoise. The slow-moving planets have qualities which must not be overlooked. The light-wave traverses the region in a few minutes and makes its report; the planet plods on and on for centuries, going over the same ground again and again. Each time it goes round it reveals a little about the space, and the knowledge slowly accumulates.

According to Newton's law a planet moves round the sun in an ellipse, and if there are no other planets disturbing it, the ellipse remains the same for ever. According to Einstein's law the path is very nearly an ellipse, but it does not quite close up; and in the next revolution the path has advanced slightly in the same direction as that in which the planet was moving. The orbit is thus an ellipse which very slowly revolves*.

The exact prediction of Einstein's law is that in one revolution of the planet the orbit will advance through a fraction of a revolution equal to $3v^2/C^2$, where v is the speed of the planet

* Appendix, Note 9.

and C the speed of light. The earth has 1/10,000 of the speed of light; thus in one revolution (one year) the point where the earth is at greatest distance from the sun will move on 3/100,000,000 of a revolution, or 0″·038. We could not detect this difference in a year, but we can let it add up for a century at least. It would then be observable but for one thing—the

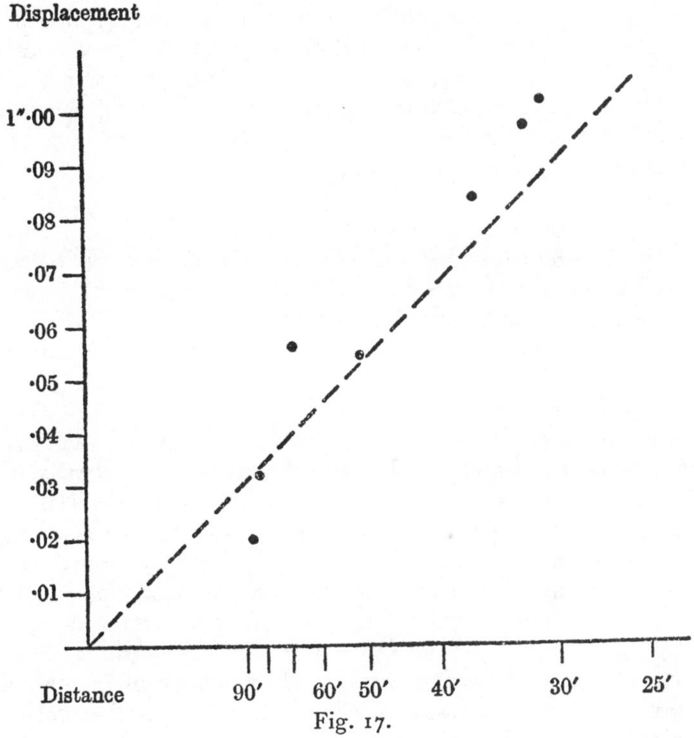

Fig. 17.

earth's orbit is very blunt, very nearly circular, and so we cannot tell accurately enough which way it is pointing and how its sharpest axes move. We can choose a planet with higher speed so that the effect is increased, not only because v^2 is increased, but because the revolutions take less time; but, what is perhaps more important, we need a planet with a sharp elliptical orbit, so that it is easy to observe how its apses move round. Both these conditions are fulfilled in the case of Mercury. It is the fastest of the planets, and the predicted advance of the orbit amounts to 43″ per century; further the eccentricity of its orbit is far greater than of any of the other seven planets.

Now an unexplained advance of the orbit of Mercury had long been known. It had occupied the attention of Le Verrier, who, having successfully predicted the planet Neptune from the disturbances of Uranus, thought that the anomalous motion of Mercury might be due to an interior planet, which was called Vulcan in anticipation. But, though thoroughly sought for, Vulcan has never turned up. Shortly before Einstein arrived at his law of gravitation, the accepted figures were as follows. The actual observed advance of the orbit was 574″ per century; the calculated perturbations produced by all the known planets amounted to 532″ per century. The excess of 42″ per century remained to be explained. Although the amount could scarcely be relied on to a second of arc, it was at least thirty times as great as the probable accidental error.

The big discrepancy from the Newtonian gravitational theory is thus in agreement with Einstein's prediction of an advance of 43″ per century....

The theory of relativity has passed in review the whole subject-matter of physics. It has unified the great laws, which by the precision of their formulation and the exactness of their application have won the proud place in human knowledge which physical science holds to-day. And yet, in regard to the nature of things, this knowledge is only an empty shell—a form of symbols. It is knowledge of structural form, and not knowledge of content. All through the physical world runs that unknown content, which must surely be the stuff of our consciousness. Here is a hint of aspects deep within the world of physics, and yet unattainable by the methods of physics. And, moreover, we have found that where science has progressed the farthest, the mind has but regained from nature that which the mind has put into nature.

We have found a strange foot-print on the shores of the unknown. We have devised profound theories, one after another, to account for its origin. At last, we have succeeded in reconstructing the creature that made the foot-print. And Lo! it is our own.

II. THE ATOMIC THEORY

LUCRETIUS

The second great problem which faces the enquirer into Nature is the structure of matter. Is matter continuous? Can it be subdivided without limit? Is water water and iron iron, if a drop of the one or a lump of the other be cut in half an infinite number of times? Or shall we come eventually to particles which, could they be still further divided, would give smaller particles no longer of water or iron but of some more fundamental stuff of which perhaps both water and iron are made?

As far as historic records go, this problem was first attacked in a systematic way by the early Ionian philosophers. They speculated about a single element, a common basis to all substances. On the other hand, Empedocles taught four primary elements, earth, water, air and fire. Leucippus and Democritus developed these concepts into a definite atomic theory, known to us best by the account of it given by the Latin Poet Lucretius.

The true meaning of the theory was its attempt to explain the different qualities of bodies in terms of rational ideas. "According to convention," says Democritus, "there is a sweet and a bitter, a hot and a cold, and according to convention there is colour. In truth there are atoms and a void." By differences in the size, shape, position and movement of atoms identical in substance, all the various kinds of matter are formed. This attempted explanation of facts by apparently simpler ideas is the essence of scientific advance. We now see that the theory of the atomists does but push back the mystery one step, while leaving it unsolved. But when such a step is made, the mind of man inevitably overestimates its importance and—a story curiously reiterated through history—discards the gods who dwell behind the mystery. The atomists of the ancient world were attacked as atheists. The French encyclopædists of the eighteenth century thought that they were not far from an explanation of the universe, and Laplace told Napoleon he had no need of the "hypothesis of God." The Inquisition condemned Galileo, the Bishops repudiated Darwin, and Huxley led a counter attack on the Bishops. Neither side has hurt the other, since their true domains are on different planes. But both have had to give up some untenable ground they had no right to occupy.

To illustrate the atomic theory of the Greeks, we give some extracts from a translation of Lucretius made by John Evelyn, the famous diarist and author of Evelyn's *Sylva*. Our extracts from the first book of the poem proclaim that "nothing from nothing comes," i.e. all things have an origin and matter is indestructible. It may seem to vanish, but it exists as minute insensible "seeds of all things" or atoms.

An ESSAY *on the* FIRST BOOK *of* T. LUCRETIUS CARUS *DE RERUM NATURA*

Interpreted and made English Verse
by J. Evelyn, Esq.

London: Printed for *Gabriel Bedle,* and *Thomas Collins,* and are to be sold at their shop at the Middle-Temple-Gate in *Fleetstreet.* 1656.

...For I of *Gods* and *Heaven* will discourse
And shew whence all things else derive their source,
Whence *Nature* doth create, augment, & cherish,
To what again resolve them when they perish.
What things in our discourse we *Matter* call,
Prolifique bodies, and the seeds of all,
Of if such terms do not the things comprise,
Prime Bodies name them, whence all other rise.
Gods in their nature of themselves subsist
'Tis certain, nor may ought their peace molest
For ever, unconcern'd with our affairs
And far remote, void of or grief or cares,
Need not our service, swim in full content,
Nor our good works accept, nor bad resent.
  Whilst sometimes human life dejected lay
On earth, under gross *superstitions* sway,
Whose head aloft from heaven seem'd t' appear
And mankind with its horrid shape did scare,
With mortal eyes to look on her that durst
Or contradict; a *Grecian* was the first;
Him nor the fame of *gods,* nor lightnings flash,
Nor threatning bruit of thundring Skies could dash,
But rather did his courage elevate,
Natures remotest doors to penetrate;
Thus did he with his vigorous wit transpierce
The flaming limits of the Universe.
All that was great his generous soul had view'd,
Whence what could be produc'd, what not be shew'd
And how each finite thing hath bounds, nor may
By any means from her fixt limits, stray:

Wherefore fond *Superstition* trampled lies
Beneath, we rear our Trophies to the Skies.
...Dark fears of mind, then banish quite away,
Not with the Sun-beams, or the light of day,
But by such *species*, as from Nature flow,
And what from right informed reason grow;
Which unto us this principle doth frame,
That *Out of nothing, nothing ever came.*
'Tis onely thus, that men are aw'd with fear,
Because such things in *Heaven* and *Earth* appear,
Of which, since they a reason cannot find
To a *celestial* Author they're assign'd.
But when we find that *nought of nought* can be,
What we pursue, we shall more clearly see,
And shew, whence all things first produced were,
And yet the *gods* still unconcerned are;
For, if of Nothing form'd, no use of Seed,
Since every sort would from all things proceed.
Men from the liquid Seas might then arise,
Fishes & *Fowl*, from Earth, *Beasts* from the Skies,
And other Cattel; *Bruits* uncertain birth
Would fill the waste, & cultivated earth.
Nor could from the same trees the same fruit spring
But all would change, & *all things all* would bring.
...Of *Nothing* then *Nothing* we must conclude
Results, but each thing is with *seed* indu'de,
From which all that's created comes to light
And clearly manifest themselves to sight.
...Add unto this, *Nature* to their first state
Doth all dissolve, nothing annihilate,
For if in all parts any thing could fail,
Death over all things would in time prevail;
Nor needed there a force to discompose
Their parts, or their strict union unloose:
But since in all eternal *Seeds* reside,
Till such a *blow* it meets, which it divides
Or else dissolves by subtle *Penetration*,
Nature preserves it whole from dissipation.
Beside those things remov'd by ages past,
If time did kill, and all their matter waste

Whence doth sweet *Venus* give to souls new birth
Through all their kinds? how should the various earth
Augment each kind with proper diet fed?
Whence flow the *Seas*? whence have free *Springs* their head?
Whence do the far extended *Rivers* rise?
And *Stars*, how are they nourish'd in the Skies?
Since length of times, and daies so many past,
All mortal *bodies* had ere this defac'd.
If then from that large tract, ought hath remain'd
From whence the sum of things has been maintain'd
Sure an *immortal nature* doth inspire
Them, nor can any thing to *nought* retire:
All from like force and cause dissolv'd would be,
Did not *eternal matter* keep it free:
And more or less them to their subjects bind,
One *touch* to them a cause of death they'd find
Had bodies no eternal permanence,
They would dissolve with the least violence:
But since the various bands of causes are
(Though *matter* permanent) *dissimilar*,
Bodies of things are safe 'till they receive
A force which may their proper thread unweave,
Nought then returns to nought, but parted fall
To *Bodies* of their prime *Originals*.
...Then nothing sure its being quite forsakes,
Since *Nature* one thing, from another makes;
Nor is there ought indeed which she supplies
Without the aid of something else that dies.
Since then I teach that *nought of nothing breeds*,
Or once produc'd, to nought again recedes,
Lest yet thou shouldst my Arguments disside
Because that *Elements* can not be spi'd
By humane eyes; behold what bodies now
In things thou canst not see, yet must allow:
First, mighty Winds, the rolling Seas incite,
Huge Vessels Wrack, and put the clouds to flight;
Rushing through fields, sometimes tall trees they crack;
And with their tearing blasts high mountains shake.
The *Seas* likewise in thund'ring billows rise
And with their raging murmur threat the Skies.

Winds therefore unseen *bodies* are, which sweep
The fleeting clouds, the Earth, the Azure deep,
Bearing with sudden storm all things away,
Yet thus proceeding, do they nought destroy
Other than as the yeelding water flowes,
Augmented by large showres, or melted snows
Wch from deep clifts in *Cataracts* descend,
Whole trees they float, and prostrate woods they rend:
Nor can strong bridges their approach sustain,
Whose rapid torrent do's all check disdain.
The River with immoderate showres repleat,
Against their *Piles* impetuously does beat:
Roaring it ruins, huge stones along it rolles,
All things it spoyles, and nothing it controles.
Even so the gusts of sturdy winds do tend
Like swiftest Rivers when they downwards bend,
And carrie all before with double might,
Sometimes they snatch, and hurry things upright
In rapid whirle. Therefore I add agen
The Winds are Bodies, and yet are not seen.
Since their effects and motions every where
Like *Rivers* be, whose *bodies* do appear.
Besides, of things we smel the various *sents*,
Which yet no substance to our sight presents;
We with our eyes see neither *Heat* nor *Cold*,
Nor can we any *Voyces* found behold
Which of *Corporeal nature* yet consist,
For they the *Sense* affect 'tis manifest.
Touch and be touch't, nought save a body may:
Cloaths become moist, wch we on shoars display;
Spread in the Sun, again, they dry appear:
But neither how that humour entred there
Can we perceive: nor by what means it flies
The heat so soon, and consequently dries.
Therefore that which is humid separates
By minute parts, which no eye penetrates.
Thus at the bare return of sundry years
The *Ring* which one upon his finger wears
Diminisheth: *Drops* which do oft distill,
Hollow hard *stones*; And whilst the field we till,

The *Coulter* of the Plough is lèssened:
And paved ways, whereon the people tread
Wear out we see: *Brass Statues* at our gates
Shew their *right hand*, wch frequent *touch* abates
Of such as visit oft, or pass the way;
Therefore things often worn the more decay:
But in each time, what *bodies* do discar'd
Is a fine sight from our gross eyes debar'd;
Lastly, what Nature by *minute* degrees
And time applies, our sharpest eye-sight flees;
Nor what through *age* or *leanness* do's decay,
Nor what from rocks at Sea *time* wears away
With gnawing salt consum'd, do we espy:
Nature with *bodies* then unseen to th' eye
All things doth manage; not that I suppose
Nature with *Bodies* do's each thing inclose
On every side, for there's a *Voyd* in things
Which rightly to conceive, much profit brings.

ALCHEMY THE PARENT OF CHEMISTRY

The philosophic and speculative mind of the Greeks did not lead to experimental science, but, as the western world emerged from the dark ages, we find first Arabians and then Europeans engaged in a curious combination of astrology, medicine and metallurgy, called the sacred art or alchemy. The belief that the stars influenced life on earth led to the known metals being named from the planets and the sun and moon. The ills to which human flesh is heir gave rise to the search for a universal remedy or *elixir vitae*, while a few crude and misunderstood experiments on metals confirmed the idea of transmutation of one element into another, especially the hope that common metals could be transmuted into nobler ones, which naturally followed from the Greek theory of a common basis of matter. Thus man's desire for wealth and power was set on an ardent hunt for a means of making gold and silver.

Alchemy, it is true, by fostering experiment, led to the birth of chemistry, but most alchemists were very far from the modern scientific attitude of mind. When they wrote of their work at all, it was in a cloud of verbiage, calculated at once to enhance their reputation for learning and to conceal the methods by which their results had been reached.

As an example of the writings of the alchemists we will take some extracts from a book of Theophrastus von Hohenheim, better known by his own epithet Paracelsus, the son of a Swiss physician, born towards the end of the fifteenth century. A man of original and self-confident mind, he soon became dissatisfied with the traditional and orthodox learning of his time, and taught only what he had himself observed in the mines of the Tyrol, at the bedside of his patients, and in the laboratory of his dispensary. He thus came to apply chemistry to medicine, as well as to study the problems of alchemy.

His writings well illustrate the characteristic confused treatment of scientific problems by the later mediaeval mind, before the Renaissance cleared the air, and definite and limited problems, suited to experimental study, came to be formulated.

(Hermetic and Alchemical Writings of Paracelsus. Translated by ARTHUR EDWARD WAITE. London 1894.)

THE COELUM PHILOSOPHORUM

By PHILIPPUS THEOPHRASTUS PARACELSUS.

THE SCIENCE AND NATURE OF ALCHEMY, AND WHAT OPINION SHOULD BE FORMED THEREOF. *Regulated by the Seven Rules or Fundamental Canons according to the seven commonly known Metals; and containing a Preface with certain Treatises and Appendices.*

THE PREFACE

OF THEOPHRASTUS PARACELSUS TO ALL ALCHEMISTS AND READERS OF THIS BOOK.

YOU who are skilled in Alchemy, and as many others as promise yourselves great riches or chiefly desire to make gold and silver, which Alchemy in different ways promises and teaches, equally, too, you who willingly undergo toil and vexations, and wish not to be freed from them, until you have attained your rewards, and the fulfilment of the promises made to you; experience teaches this every day, that out of thousands of you not even one accomplishes his desire. Is this a failure of Nature or of Art? I say no; but it is rather the fault of fate, or of the unskilfulness of the operator.

Since, therefore, the characters of the signs, of the stars and planets of heaven, together with the other names, inverted words, receipts, materials, and instruments are thoroughly well known to such as are acquainted with this art, it would be altogether superfluous to recur to these same subjects in the present book, although the use of such signs, names, and characters at the proper time is by no means without advantage.

But herein will be noticed another way of treating Alchemy different from the previous method, and deduced by Seven Canons from the sevenfold series of the metals. This, indeed, will not give scope for a pompous parade of words, but, nevertheless, in the consideration of those Canons everything which should be separated from Alchemy will be treated at sufficient length, and, moreover, many secrets of other things are herein contained. Hence, too, result certain marvellous speculations and new operations which frequently differ from the writings and opinions of ancient operators and natural philosophers, but have been discovered and confirmed by full proof and experimentation.

Moreover, in this Art nothing is more true than this, though it be little known and gains small confidence. All the fault and cause of difficulty in Alchemy, whereby very many persons are reduced to poverty, and others labour in vain, is wholly and solely lack of skill in the operator, and the defect or excess of materials, whether in quantity or quality, whence it ensues that, in the course of operation, things are wasted or reduced to nothing. If the true process shall have been found, the substance itself while transmuting approaches daily more and more towards perfection. The straight road is easy, but it is found by very few.

Sometimes it may happen that a speculative artist may, by his own eccentricity, think out for himself some new method in Alchemy, be the consequence anything or nothing. He need do nought in order to reduce something into nothing, and again bring back something out of nothing. Yet this proverb of the incredulous is not wholly false. Destruction perfects that which is good; for the good cannot appear on account of that which conceals it. The good is least good while it is thus concealed. The concealment must be removed so that the good may be able freely to appear in its own brightness. For example, the mountain, the sand, the earth, or the stone in which a metal has grown is such a concealment. Each one of the visible metals is a concealment of the other six metals.

By the element of fire all that is imperfect is destroyed and taken away, as, for instance, the five metals, Mercury, Jupiter, Mars, Venus, and Saturn. On the other hand, the perfect metals, Sol and Luna, are not consumed in that same fire. They remain in the fire: and at the same time, out of the other imperfect ones which are destroyed, they assume their own body and become visible to the eyes. How, and by what method, this comes about can be gathered from the Seven Canons. Hence it may be learnt what are the nature and property of each metal, what it effects with the other metals, and what are its powers in commixture with them.

But this should be noted in the very first place: that these Seven Canons cannot be perfectly understood by every cursory reader at a first glance or a single reading. An inferior intelligence does not easily perceive occult and abstruse subjects. Each one of these Canons demands no slight discussion. Many persons, puffed up with pride, fancy they can easily comprehend all which this book comprises. Thus they set down its contents as useless and futile, thinking they have something far better of their own, and that therefore they can afford to despise what is here contained.

THE SEVEN CANONS OF THE METALS

THE FIRST CANON. CONCERNING THE NATURE AND PROPERTIES OF MERCURY

...By the mediation of Vulcan, or fire, any metal can be generated from Mercury. At the same time, Mercury is imperfect as a metal; it is semi-generated and wanting in coagulation, which is the end of all metals. Up to the half-way point of their generation all metals are Mercury. Gold, for example is Mercury; but it loses the Mercurial nature by coagulation, and although the properties of Mercury are present in it, they are dead, for their vitality is destroyed by coagulation....

THE SECOND CANON. CONCERNING THE NATURE AND PROPERTIES OF JUPITER (*tin*).

In that which is manifest (that is to say, the body of Jupiter) the other six corporeal metals are spiritually concealed, but one more deeply and more tenaciously than another. Jupiter has nothing of a Quintessence in his composition, but is of the nature of the four elementaries. On this account his liquefaction is brought about by the application of a moderate fire, and, in like manner, he is coagulated by moderate cold. He has affinity with the liquefactions of all the other metals. For the more like he is to some nature, the more easily he is united thereto by conjunction. For the operation of those nearly allied is easier and more natural than of those which are remote....The more remote, therefore, Jupiter is found to be from Mars and Venus, and the nearer Sol and Luna, the more "goldness" or "silveriness," if I may say so, it contains in his body, and the greater, stronger, more visible, more tangible, more amiable, more acceptable, more distinguished, and more true it is found than in some remote body....This, therefore, is a point which you, as an Alchemist, must seriously debate with yourself, how you can relegate Jupiter to a remote and abstruse place, which Sol and Luna occupy, and how, in turn, you can summon Sol and Luna from remote positions to a near place, where Jupiter is corporeally posited; so that, in the same way, Sol and Luna may also be present there corporeally before your eyes. For the transmutation of metals from imperfection to perfection there are several practical receipts. Mix the one with the other. Then again separate the one pure from the other. This is nothing else but the process of permutation, set in order by perfect alchemical labour. Note that Jupiter has much gold and not a little silver. Let Saturn and Luna be imposed on him, and of the rest Luna will be augmented.

THE THIRD CANON. CONCERNING MARS (*iron*) AND HIS PROPERTIES

The six occult metals have expelled the seventh from them, and have made it corporeal, leaving it little efficacy, and imposing on it great hardness and weight. This being the case, they have shaken off all their own strength of coagulation and hardness, which they manifest in this other body. On the contrary, they have retained in themselves their colour and liquefaction, together with their nobility. It is very difficult and laborious for a prince or a king to be produced out of an unfit and common man. But Mars acquires dominion with strong and pugnacious hand, and seizes on the position of king. He should, however, be on his guard against snares, that he be not led captive suddenly and unexpectedly. It must also be considered by what method Mars may be able to take the place of king, and Sol and Luna, with Saturn, hold the place of Mars.

THE FOURTH CANON. CONCERNING VENUS (*copper*) AND ITS PROPERTIES

The other six metals have rendered Venus an extrinsical body by means of all their colour and method of liquefaction. It may be necessary, in order to understand this, that we should show, by some examples, how a manifest thing may be rendered occult, and an occult thing rendered materially manifest by means of fire. Whatever is combustible can be naturally transmuted by fire from one form into another, namely, into lime, soot, ashes, glass, colours, stones, and earth. This last can again be reduced to many new metallic bodies. If a metal, too, be burnt, or rendered fragile by old rust, it can again acquire malleability by applications of fire.

THE FIFTH CANON. CONCERNING THE NATURE AND PROPERTIES OF SATURN (*lead*).

Of his own nature Saturn speaks thus: the other six have cast me out as their examiner. They have thrust me forth from them and from a spiritual place. They have also added a corruptible body as a place of abode, so that I may be what they neither are nor desire to become. My six brothers are spiritual, and thence it ensues that so often as I am put in the fire they penetrate my body and, together with me, perish in the fire, Sol and Luna excepted...

THE SIXTH CANON. CONCERNING LUNA (*silver*) AND THE PROPERTIES THEREOF

The endeavour to make Saturn or Mars out of Luna involves no lighter or easier work than to make Luna, with great gain, out of Mercury, Jupiter, Mars, Venus, or Saturn. It is not useful to transmute what is perfect into what is imperfect, but the latter into the former. Nevertheless, it is well to know what is the material of Luna, or whence it proceeds. Whoever is not able to consider or find this out will neither be able to make Luna. It will be asked, What is Luna? It is among the seven metals which are spiritually concealed, itself the seventh, external, corporeal, and material. For this seventh always contains the six metals spiritually hidden in itself. And the six spiritual metals do not exist without one external and material metal. So also no corporeal metal can have place or essence without those six spiritual ones. The seven corporeal metals mix easily by means of liquefaction, but this mixture is not useful for making Sol or Luna. For in that mixture each metal remains in its own nature, or fixed in the fire, or flies from it. For example, mix, in any way you can, Mercury, Jupiter, Saturn, Mars, Venus, Sol, and Luna. It will not thence result that Sol and Luna will so change the other five that, by the agency of Sol and Luna, these will become Sol and Luna. For though all be liquefied into a single mass, nevertheless each remains in its nature whatever it is. This is the judgement which must be passed on corporeal mixture....

A question may arise: If it be true that Luna and every metal derives its origin and is generated from the other six, what is then its property and its nature? To this we reply: From Saturn, Mercury, Jupiter, Mars, Venus, and Sol, nothing and no other metal than Luna could be made. The cause is that each metal has two good virtues of the other six, of which altogether there are twelve. These are the spirit of Luna, which thus in a few words may be made known. Luna is composed of the six spiritual metals and their virtues, whereof each possesses two. Altogether, therefore, twelve are thus posited in one corporeal metal,...Luna has from the planet Mercury...its liquidity and bright white colour. So Luna has from Jupiter... its white colour and its great firmness in fire. Luna has from Mars...its hardness and its clear sound. Luna has from Venus ...its measure of coagulation and its malleability. From Saturn... its homogeneous body, with gravity. From Sol...its spotless purity and great constancy against the power of fire. Such is the knowledge of the natural exaltation and of the course of the spirit and body of Luna, with its composite nature and wisdom briefly summarised....

THE SEVENTH CANON. CONCERNING THE NATURE OF SOL (*gold*) AND ITS PROPERTIES

The seventh after the six spiritual metals is corporeally Sol, which in itself is nothing but pure fire. What in outward appearance is more beautiful, more brilliant, more clear and perceptible, a heavier, colder, or more homogeneous body to see? And it is easy to perceive the cause of this, namely, that it contains in itself the congelations of the other six metals, out of which it is made externally into one most compact body....The fire of Sol is of itself pure, not indeed alive, but hard, and so far shews the colour of sulphur in that yellow and red are mixed therein in due proportion. The five cold metals are Jupiter, Mars, Saturn, Venus, and Luna, which assign to Sol their virtues; according to cold, the body itself; according to fire, colour; according to dryness, solidity; according to humidity, weight; and out of brightness, sound. But that gold is not burned in the element of terrestrial fire, nor is even corrupted, is effected by the firmness of Sol. For one fire cannot burn another, or even consume it; but rather if fire be added to fire it is increased, and becomes more powerful in its operations. The celestial fire which flows to us on the earth from the Sun is not such a fire as there is in heaven, neither is it like that which exists upon the earth, but that celestial fire with us is cold and congealed, and it is the body of the Sun. Wherefore the Sun can in no way be overcome by our fire. This only happens, that it is liquefied, like snow or ice, by that same celestial Sun. Fire, therefore, has not the power of burning fire, because the Sun is fire, which, dissolved in heaven, is coagulated with us.

The End of the Seven Canons

CERTAIN TREATISES AND APPENDICES ARISING OUT OF THE SEVEN CANONS

WHAT IS TO BE THOUGHT CONCERNING THE CONGELATION OF MERCURY

To mortify or congeal Mercury, and afterwards seek to turn it into Luna, and to sublimate it with great labour, is labour in vain, since it involves a dissipation of Sol and Luna existing therein. There is another method, far different and much more concise, whereby, with little waste of Mercury and less expenditure of toil, it is transmuted into Luna without congelation. Any one can at pleasure learn this art in Alchemy, since it is so simple and easy; and by it, in a short time, he could make any quantity of silver and gold. It is tedious to read long descriptions, and everybody wishes to be advised in straightforward words. Do this, then; proceed as follows, and you will have Sol and Luna, by help whereof you will turn out a very rich man. Wait awhile, I beg, while this process is described to you in a few words, and keep these words well digested, so that out of Saturn, Mercury, and Jupiter you may make Sol and Luna. There is not, nor ever will be, any art so easy to find out and practise, and so effective in itself. The method of making Sol and Luna by Alchemy is so prompt that there is no more need of books, or of elaborate instruction, than there would be if one wished to write about last year's snow.

Concerning the Receipts of Alchemy

What, then, shall we say about the receipts of Alchemy, and about the diversity of its vessels and instruments? These are furnaces, glasses, jars, waters, oils, limes, sulphurs, salts, saltpetres, alums, vitriols, chrysocollæ, copper-greens, atraments, auri-pigments, fel vitri, ceruse, red earth, thucia, wax, lutum sapientiæ, pounded glass, verdigris, soot, crocus of Mars, soap, crystal, arsenic, antimony, minium, elixir, lazarium, gold-leaf, salt-nitre, sal ammoniac, calamine stone, magnesia, bolus armenus, and many other things. Moreover, concerning preparations, putrefactions, digestions, probations, solutions, cementings, filtrations, reverberations, calcinations, graduations, rectifications, amalgamations, purgations, etc., with these alchemical books are crammed. Then, again, concerning herbs, roots, seeds, woods, stones, animals, worms, bone dust, snail shells, other shells, and pitch. These and the like, whereof there are some very far-fetched in Alchemy, are mere incumbrances of work; since even if Sol and Luna could be made by them they rather hinder and delay than further one's purpose. But it is not from these—to say the truth—that the Art of making Sol and Luna is to be learnt. So, then, all these things should be passed by, because they have no effect with the five metals, so far as Sol and Luna are concerned. Someone may ask, What, then, is this short and easy way, which involves no difficulty, and yet whereby Sol and Luna can be made? Our answer is, this has been fully and openly explained in the Seven Canons. It would be lost labour should one seek further to instruct one who does not understand these. It would be impossible to convince such a person that these matters could be so easily understood, but in an occult rather than in an open sense.

THE ART IS THIS; After you have made heaven, or the sphere of Saturn, with its life to run over the earth, place it on all the planets, or such, one or more, as you wish, so that the portion of Luna may be the smallest. Let all run, until heaven, or Saturn, has entirely disappeared. Then all those planets will remain dead with their old corruptible bodies, having meanwhile obtained another new, perfect and incorruptible body.

That body is the spirit of heaven. From it these planets again receive a body and life, and live as before. Take this body from the life and the earth. Keep it. It is Sol and Luna. Here you have the Art altogether, clear and entire. If you do not yet understand it, or are not practised therein, it is well. It is better that it should be kept concealed, and not made public.

LAVOISIER AND THE RISE OF MODERN CHEMISTRY

MODERN chemistry began with the successful explanation of combustion and respiration. The wonderful phenomena of flame have always fascinated the human mind. To some of the Greeks, as we have seen, fire was the noblest of the elements. To the men who followed the Renaissance, it seemed that flame escaped when a body burned, and, since the calx which sometimes remained was heavier than the fuel, the escaping essence must have a negative weight. We thus return, out of due time, to Aristotle's idea of a substance essentially light in nature. To this substance the name of "phlogiston" was given by Stahl (1660–1734), who considered it to be one of the elements.

Within a few years, the experiments of Black, Cavendish and Priestley gave enough evidence to overthrow Stahl's theory, but it was Antoine Laurent Lavoisier (1743–1794), who grasped their significance and added new confirmatory work of his own. Lavoisier saw that there was no need to invent a body with properties unlike those of common substances; that combustion was but intense chemical action between the fuel and an active gas forming part of the air which was also used by animals when they breathe. To this gas he afterwards gave the name of oxygen. Thus, as Galileo and Newton proved that terrestrial mechanics held good in the heavens, so Lavoisier showed that the mysterious phenomena of flame and the breath of life itself could be brought into conformity with ordinary chemistry, and subjected to exact methods of analysis. Lavoisier, a man of affairs and a master of all the science of his time, had the honour of becoming obnoxious to the Jacobins of the French Revolution, and was sent to the guillotine to prove that "the republic has no need of savants."

WORKS OF LAVOISIER

Memoir on the Nature of the Principle which Combines with Metals during Calcination and Increases their Weight

(*Read to the Académie des Sciences, Easter*, 1775.)

Are there different kinds of air?...Are the different airs that nature offers us, or that we succeed in making, exceptional substances, or are they modifications of atmospheric air? Such are the principal subjects embraced in my scheme of work, the development of which I propose to submit to the Academy. But since the time devoted to our public meetings does not allow me to treat any one of these questions in full, I will confine myself to-day to one particular case, and will only show that the principle which unites with metals during calcination is nothing else than the healthiest and purest part of air, so that if air, after entering into combination with a metal, is set free again, it emerges in an eminently respirable condition, more suited than atmospheric air to support ignition and combustion.

The majority of metallic calces are only reduced, that is to say, only return to the metallic condition, by immediate contact with a carbonaceous material, or with some substance containing what is called *phlogiston*. The charcoal that one uses is entirely destroyed during the operation when the amount is in suitable proportion, whence it follows that the air set free from metallic reductions with charcoal is not simple; it is in some way the result of the combination of the elastic fluid set free from the metal and that set free from the charcoal; thus, though this fluid is obtained in the state of fixed air it is not justifiable to conclude that it existed in this state in the metallic calx before its combination with the carbon.

These reflections made me feel how essential it was—in order to unravel the mystery of the reduction of metallic calces —to perform all my experiments on calces which can be reduced without addition of charcoal....

Precipitated mercury, which is nothing else than a calx of mercury, as several authors have suggested and as this memoir will further show, precipitated mercury, I repeat, seemed to me suitable for the object I had in view: for nobody to-day is unaware that this substance can be reduced without addition of charcoal at a very moderate degree of heat. Although I have repeated many times the experiments I am going to quote, I have not thought it suitable to give particular details of any of them here, for fear of occupying too much space, and I have therefore combined into one account the observations made during several repetitions of the same experiment.

In order to be sure that precipitated mercury was a true metallic calx, that it gave the usual results and the usual kind of air on reduction by the ordinary method, that is to say, using the recognized expression, by the addition of phlogiston, I mixed one ounce of this calx with 48 grains of powdered charcoal, and introduced the whole into a little glass retort of at most two cubic inches capacity, which I placed in a reverberatory furnace of proportionate size. The neck of this retort was about a foot long, and three to four lines in diameter; it had been bent in a flame in different places and its tip was such that it could be fixed under a bell-jar of sufficient size, filled with water, and turned upside down in a trough of water....

As soon as a flame was applied to the retort and the heat had begun to take effect, the ordinary air contained in the retort expanded, and a small quantity passed into the bell-jar; but in view of the small size of the part of the retort that remained empty, this air could not introduce a sensible error, and at the most it could scarcely amount to a cubic inch. When the retort began to get hotter, air was very rapidly evolved and bubbled up through the water into the bell-jar; the operation did not last more than three-quarters of an hour, the flame being used sparingly during this interval. When the calx of mercury was reduced and air was no longer evolved, I marked the height at which the water stood in the bell-jar and found that the air set free amounted to 64 cubic inches, without allowing for the volume necessarily dissolved in the water.

I submitted this air to a large number of tests, which I will not describe in detail, and found (1) that it combined with water on shaking and gave to it all the properties of acidulated, gaseous, or aerated waters such as those of Seltz, Bougues, Bussang, Pyrmont, etc.; (2) that animals placed in it died in a few seconds; (3) that candles and all burning bodies were instantly extinguished therein; (4) that it precipitated lime water; (5) that it combined very easily with fixed or volatile alkalis, removing their causticity and giving them the power of crystallizing. All these are precisely the qualities of the kind of air known as *fixed air*, such as I obtained by the reduction of *minium* by powdered charcoal, such as is set free from calcareous earths and effervescent alkalis by their combination with acids, or from fermenting vegetable matters, etc. It was thus certain that precipitated mercury gave the same products as other metallic calces on reduction in the presence of phlogiston and that it could consequently be included in the general category of metallic calces.

It remained to examine this calx alone, to reduce it without addition, to see if some elastic fluid was still set free, and if so, to determine the nature of such fluid. With this in view, I put into a retort of the same size as before (two cubic inches) one ounce of precipitated mercury alone; I arranged the apparatus in the same way as for the preceding experiment, so that all the circumstances were exactly the same; the reduction was a little harder to bring about than when charcoal was present; it required more heat and there was no perceptible effect till the retort began to get slightly red-hot; then air was set free little by little, and passed into the bell-jar, and by keeping up the same degree of heat for 2½ hours, all the mercury was reduced.

The operation completed, I found 7 gros, 18 grains of liquid mercury, some in the neck of the retort and some in a glass vessel I had placed at the tip of the retort under the water; the amount of air in the bell-jar was found to be 78 cubic inches; from this it follows, if the loss of weight of the mercury is attributed to the loss of this air, that each cubic inch must weigh a little less than two-thirds of a grain, which does not differ much from the weight of ordinary air.

Having established these results, I hastened to submit the 78 cubic inches of air I had obtained to all the tests which could indicate its nature, and I found, much to my surprise:

(1) that it did not combine with water on shaking;

(2) that it did not precipitate lime water, but only caused in it an almost imperceptible turbidity;

(3) that it entered into no compounds with fixed or volatile alkalis;

(4) that it did not in the least diminish their causticity;

(5) that it could be used again for the calcination of metals;

(6) in short, that it had none of the properties of fixed air: far from causing animals to perish like fixed air, it seemed on the contrary more suited to support their respiration; not only were candles and burning objects not extinguished, but the flame increased in a very remarkable manner: it gave much more light than in common air; charcoal burned with a flash almost like that of phosphorus, and all combustible bodies were consumed with astonishing speed. All these circumstances fully convinced me that this air, far from being fixed air, was in a more respirable and combustible, and therefore in a purer condition, than even the air in which we live.

This seems to prove that the principle which unites with metals when they are calcined and causes them to increase in weight is nothing else than the purest part of the air which surrounds us, which we breathe, and which during calcination passes from a condition of expansibility to that of solidity; if it is obtained in the form of fixed air from metallic reductions in which charcoal is employed, this is due to the combination of the charcoal with the pure part of the air, and it is very probable that all metallic calces would give, like that of mercury, only this eminently respirable air, if one could reduce them all without addition, as precipitated mercury is reduced....

Experiments on the Respiration of Animals and on the Changes which happen to Air in its Passage through their Lungs

(*Read to the Académie des Sciences, 3rd May,* 1777.)

Of all the phenomena of animal economy, none are more striking, nor more worthy of attention from physicists and physiologists than those accompanying respiration. If, on the one hand, we know little of the object of this singular function, we know, on the other hand, that it is so essential to life that it cannot be suspended for any time without exposing the animal to danger of immediate death.

Air, as everyone knows, is the agent or more exactly the subject of respiration, but, at the same time, all kinds of air, or more generally all kinds of elastic fluids, are not adapted to maintain it, and there are a large number of airs that animals cannot breathe without perishing at least as promptly as if they did not breathe at all.

The experiments of certain physicists and above all those of MM. Hale and Cigna first began to cast some light on this important matter; since then, M. Priestley, in a work published in London last year, has pushed back the limits of our knowledge yet further, and he has sought to prove, by very ingenious, very delicate, and very novel experiments, that the respiration of animals has the property of phlogisticating the air, like the calcination of metals and several other chemical processes, and that air only ceases to be respirable at the moment when it is overcharged and in some way saturated with phlogiston.

However probable the theory of the celebrated physicist may have appeared at first sight, however numerous and well-performed the experiments on which he sought to establish his theory, I must own that I have found it to be contradicted by so many phenomena that I have thought myself justified in calling it in question; I have consequently worked on another plan and I have been irresistibly led by my experiments to conclusions quite other than his. I will not stop to discuss in detail each of M. Priestley's experiments, nor to demonstrate how they all prove to be in favour of the opinion I am going to expound in this memoir; I will content myself with quoting those which are necessary for my purpose....

I enclosed 50 cubic inches of common air in a suitable apparatus, which it would be difficult to describe without illustrations; I introduced into this apparatus 4 ounces of very pure mercury and proceeded to calcine it, keeping it for twelve days at a heat almost sufficient to make it boil.

Nothing noteworthy happened during the first day; the mercury, though not boiling, was in a state of continual evaporation....On the second day, red specks began to appear on the surface of the mercury, and they increased in size and volume daily; finally, at the end of twelve days, having let the fire out and allowed the vessel to cool, I observed that the air it had contained was diminished by 8 to 9 cubic inches, that is to say, by about a sixth of its volume; at the same time there had been formed a considerable quantity, approximately 45 grains, of precipitated mercury, or calx of mercury.

The air thus diminished in volume did not precipitate lime water, but it extinguished flames, caused animals placed in it to die in a short while, gave practically no red vapours with nitrous air, nor was perceptibly diminished by it—in a word, it was in an absolutely mephitic condition.

The experiments of M. Priestley and myself have made clear that precipitated mercury is nothing else than a compound of mercury with about one-twelfth of its own weight of a kind of air that is better and more respirable, if I may use such a word, than common air; it appears then to be proved that in the foregoing experiment the mercury absorbed the better and more respirable part of the air during calcination, leaving behind only the mephitic or non-respirable part: the following experiment confirmed this truth yet further.

I carefully collected the 45 grains of calx of mercury formed in the preceding calcination; I put them in a very small glass retort whose neck, doubly bent, was fixed under a bell-jar full of water, and I proceeded to reduce it without adding anything.

By this operation I recovered almost the same amount of air that had been absorbed by the calcination, that is to say, 8 to 9 cubic inches, and on combining these 8 to 9 cubic inches with the air vitiated by the calcination of mercury, I restored this air exactly enough to its state before calcination, i.e., to the state of common air: the air thus restored no longer extinguished flames, no longer caused the death of animals breathing it, and finally was almost as much diminished by nitrous air as the air of the atmosphere.

Here is the most complete kind of proof to which one can attain in chemistry, the decomposition and recomposition of air, and from this one may evidently conclude (1) that five-sixths of the air we breathe are incapable of supporting the respiration of animals, or ignition and combustion; (2) that the surplus, i.e., one fifth only of the volume of atmospheric air is respirable; (3) that when mercury is calcined, it absorbs the healthy part of the air, leaving only the noxious part; (4) that on bringing together the two parts of the air thus separated, the respirable part and the noxious part, one reproduces air similar to that of the atmosphere.

The preliminary facts concerning the calcination of metals lead us to simple conclusions concerning the respiration of animals, and as air which has for some time served to support this vital function has much in common with that in which metals have been calcined, knowledge of the one may naturally be applied to the other....

Air which had been breathed by an animal till it died became very different from atmospheric air; it precipitated lime water; it extinguished flames; it was no longer diminished by nitrous air; another animal introduced into it only lived some seconds; it was entirely mephitic, and in this respect appeared similar enough to that which remained after the calcination of mercury.

However a more thorough examination showed me two very remarkable differences between these two airs, I mean between that which had been used for the calcination of mercury and that which had been used to support respiration: firstly, the diminution of volume was much less in the latter than in the former; secondly, air from respiration precipitated lime water, while air from calcination caused no change therein.

This difference, on the one hand, between these two airs, and, on the other hand, the close analogy which they showed in many respects, led me to assume that two processes were confused in respiration of which I probably yet knew but one, and to remove my uncertainties on this matter, I made the following experiment.

I caused 12 cubic inches of air vitiated by respiration to pass into a bell-jar full of mercury, turned upside down in a trough full of mercury, and I introduced also a thin layer of fixed caustic alkali; I might have used lime water instead, but the volume which it would have been necessary to employ would have been too large and would have been prejudicial to the success of the experiment.

The effect of the caustic alkali was to occasion a diminution of nearly one-sixth in the volume of the air; at the same time the alkali partly lost its causticity; it acquired the property of effervescing with acids, and crystallized even under the bell-jar in very regular rhomboids; properties which one knows cannot be communicated to it except by its combination with the kind of air or gas known as *fixed air*, which I shall henceforward call "*gaseous acid of chalk*"; it follows that air vitiated by respiration contains almost one-sixth of a gaseous acid exactly similar to that obtained from chalk.

Far from being restored to the state of common air, the air thus freed from its fixable portion by caustic alkali had more resemblance to air used for the calcination of mercury, or rather was one and the same thing; like the latter, it caused animals to die, extinguished flames; in fact, none of the experiments I made to compare these two airs enabled me to perceive the slightest difference between them.

But air which has been used for the calcination of mercury is nothing else, as we have seen above, than the mephitic residue of atmospheric air, the respirable part having combined with the mercury during the calcination; so that air which has been used for respiration, when it has been deprived of the gaseous acid of chalk that it contains, is likewise nothing but the residue of common air deprived of its respirable portion; and, in fact, having added to this air about a quarter of its volume of respirable air from calx of mercury, I restored it to its original condition and made it as suitable as ordinary air for respiration, or for the support of combustion in the same way as I had restored air which had been vitiated by the calcination of metals.

It follows from these experiments that to change air vitiated by respiration back to the state of ordinary respirable air, it is necessary (1) to remove from this air by means of lime or a caustic alkali the portion of gaseous acid of chalk that it contains; (2) to restore to it a quantity of respirable or dephlogisticated air equal to that which it has lost. Respiration, by a necessary consequence, must effect the reverse of these two changes, and I find myself led to two equally probable explanations of this, between which experiment does not yet allow us to decide.

For from what goes above one may conclude that one of two things occurs as a result of respiration: either the respirable air contained in the air of the atmosphere is converted into gaseous acid of chalk during its passage through the lung; or there is an exchange in this organ, respirable air being absorbed and an almost equal volume of gaseous acid of chalk being given up by the lung....

Memoir on the Combustion of Candles in Atmospheric Air and in Respirable Air

(*Communicated to the Académie des Sciences*, 1777.)

I have established in the foregoing memoirs that the air of the atmosphere is not a simple substance, an element, as the ancients believed and as has been supposed till our time; that the air we breathe is composed of respirable air to the extent of only one quarter and that the remainder is a noxious gas (probably itself of a complex nature) which cannot alone support the life of animals, or combustion or ignition. I feel obliged, consequently, to distinguish four kinds of air or air-like fluids.

Firstly, atmospheric air; that in which we live, which we breathe, etc.

Secondly, pure air, respirable air; that which forms only a quarter of atmospheric air and which M. Priestley has very wrongly called *dephlogisticated air*.

Thirdly, the noxious air which makes up three quarters of atmospheric air and whose nature is still entirely unknown to us.

Fourthly, fixed air, which I shall call henceforward, following M. Bucquet, by the name of *acid of chalk*.

CHEMISTRY AND THE ATOMIC THEORY

By experiments on falling bodies, Galileo disproved Aristotle's conception of bodies intrinsically light, and thus made possible a consistent theory of dynamics. The corresponding simplification in chemistry was made when Lavoisier impressed on the world the fact that there was no need to explain combustion by the conception of a body with properties fundamentally unlike those of other material substances. The unity of outlook which followed made modern chemistry possible. Many facts already known fell into a reasonable scheme of knowledge, and innumerable fresh facts came to light under the impetus given to chemistry by the new and simpler views.

It was inevitable that the fundamental question of the nature of matter should come up for reconsideration. The atomic theory of the Greeks, which we have seen mirrored in the poetry of Lucretius, was held as a probable speculation by many seventeenth and eighteenth century philosophers, and used by physicists such as Newton and Boyle. But no real and considerable advance was made in the theory itself till Lavoisier's work had simplified and defined the problem, and the development of experimental chemistry had given firm ground from which to start. The theory then ceased to be a philosophic speculation, and became a definite scientific hypothesis, framed to explain certain quantitative measurements, and itself of a nature to be put into a quantitative form. In other words, the measurement of the relative weights in which chemical elements combine gave the relative weights of their constituent atoms. This result is due to John Dalton (1766–1844), who was born of a Quaker family in Cumberland, and worked and died at Manchester.

A NEW SYSTEM OF CHEMICAL PHILOSOPHY

By John Dalton, 1808.

Part I. Chap. II. *On the Constitution of Bodies.*

There are three distinctions in the kinds of bodies, or three states, which have more especially claimed the attention of philosophical chemists; namely, those which are marked by the terms *elastic fluids, liquids, and solids.* A very famous instance is exhibited to us in water, of a body, which, in certain circumstances, is capable of assuming all the three states. In steam we recognise a perfectly elastic fluid, in water a perfect liquid, and in ice a complete solid. These observations have tacitly led to the conclusion which seems universally adopted, that all bodies of sensible magnitude, whether liquid or solid, are constituted of a vast number of extremely small particles, or atoms of matter bound together by a force of attraction, which is more or less powerful according to circumstances....

Whether the ultimate particles of a body, such as water, are all alike, that is, of the same figure, weight, &c. is a question of some importance. From what is known, we have no reason to apprehend a diversity in these particulars: if it does exist in water, it must equally exist in the elements constituting water, namely, hydrogen and oxygen. Now it is scarcely possible to conceive how the aggregates of dissimilar particles should be so uniformly the same. If some of the particles of water were heavier than others, if a parcel of the liquid on any occasion were constituted principally of these heavier particles, it must be supposed to affect the specific gravity of the mass, a circumstance not known. Similar observations may be made on other substances. Therefore we may conclude that *the ultimate particles of all homogeneous bodies are perfectly alike in weight, figure, &c.* In other words, every particle of water is like every other particle of water; every particle of hydrogen is like every other particle of hydrogen, &c.

Part I. Chap III. *On Chemical Synthesis.*

When any body exists in the elastic state, its ultimate particles are separated from each other to a much greater distance than in any other state; each particle occupies the centre of a comparatively large sphere, and supports its dignity by keeping all the rest, which by their gravity, or otherwise, are disposed to encroach upon it, at a respectful distance. When we attempt to conceive the *number* of particles in an atmosphere, it is somewhat like attempting to conceive the number of stars in the universe: we are confounded with the thought. But if we limit the subject, by taking a given volume of any gas, we seem persuaded that, let the divisions be ever so minute, the number of particles must be finite; just as in a given space of the universe, the number of stars and planets cannot be infinite.

Chemical analysis and synthesis go no farther than to the separation of particles one from another, and to their reunion. No new creation or destruction of matter is within the reach of chemical agency. We might as well attempt to introduce a new planet into the solar system, or to annihilate one already in existence, as to create or destroy a particle of hydrogen. All the changes we can produce, consist in separating particles that are in a state of cohesion or combination, and joining those that were previously at a distance.

In all chemical investigations, it has justly been considered an important object to ascertain the relative *weights* of the simples which constitute a compound. But unfortunately the enquiry has terminated here; whereas from the relative weights in the mass, the relative weights of the ultimate particles or atoms of the bodies might have been inferred, from which their number and weight in various other compounds would appear, in order to assist and to guide future investigations, and to correct their results. Now it is one great object of this work, to shew the importance and advantage of ascertaining *the relative weights of the ultimate particles, both of simple and compound bodies, the number of simple elementary particles which constitute one compound particle, and the number of less compound particles which enter into the formation of one more compound particle.*

If there are two bodies, *A* and *B*, which are disposed to combine, the following is the order in which the combinations may take place, beginning with the most simple: namely,

1 atom of A + 1 atom of B = 1 atom of C, binary.

1 atom of A + 2 atoms of B = 1 atom of D, ternary.

2 atoms of A + 1 atom of B = 1 atom of E, ternary.

1 atom of A + 3 atoms of B = 1 atom of F, quaternary.

3 atoms of A + 1 atom of B = 1 atom of G, quaternary.

&c. &c.

The following general rules may be adopted as guides in all our investigations respecting chemical synthesis.

1st. When only one combination of two bodies can be obtained, it must be presumed to be a *binary* one, unless some cause appear to the contrary.

2d. When two combinations are observed, they must be presumed to be a *binary* and a *ternary*.

3d. When three combinations are obtained, we may expect one to be a *binary*, and the other two *ternary*.

4th. When four combinations are observed, we should expect one *binary*, two *ternary*, and one *quaternary*, &c.

5th. A *binary* compound should always be specifically heavier than the mere mixture of its two ingredients.

6th. A *ternary* compound should be specifically heavier than the mixture of a binary and a simple, which would, if combined, constitute it; &c.

7th. The above rules and observations equally apply, when two bodies, such as *G* and *D*, *D* and *E*, &c. are combined.

From the application of these rules, to the chemical facts already well ascertained, we deduce the following conclusions: 1st. That water is a binary compound of hydrogen and oxygen, and the relative weights of the two elementary atoms are as 1 : 7, nearly; 2d. That ammonia is a binary compound of hydrogen and azote, and the relative weights of the two atoms are as 1 : 5, nearly; 3d. That nitrous gas is a binary compound of azote and oxygen, the atoms of which weigh 5 and 7 respectively; that nitric acid is a binary or ternary compound according as it is derived, and consists of one atom of azote and two of oxygen, together weighing 19; that nitrous oxide is a compound similar to nitric acid, and consists of one atom of oxygen and two of azote, weighing 17; that nitrous acid is a binary compound of nitric acid and nitrous gas, weighing 31; that oxynitric acid is a binary compound of nitric acid and oxygen, weighing 26; 4th. That carbonic oxide is a binary compound, consisting of one atom of charcoal, and one of oxygen, together weighing nearly 12; that carbonic acid is a ternary compound, (but sometimes binary) consisting of one atom of charcoal, and two of oxygen, weighing 19; &c. &c. In all these cases the weights are expressed in atoms of hydrogen, each of which is denoted by unity.

In the sequel, the facts and experiments from which these conclusions are derived, will be detailed; as well as a great variety of others from which are inferred the constitution and weight of the ultimate particles of the principal acids, the alkalis, the earths, the metals, the metallic oxides and sulphurets, the long train of neutral salts, and in short, all the chemical compounds which have hitherto obtained a tolerably good analysis. Several of the conclusions will be supported by original experiments.

From the novelty as well as importance of the ideas suggested in this chapter, it is deemed expedient to give plates, exhibiting the mode of combination in some of the more simple cases. A specimen of these accompanies this first part. The elements or atoms of such bodies as are conceived at present to be simple, are denoted by a small circle, with some distinctive mark; and the combinations consist in the juxtaposition of two or more of these; when three or more particles of elastic fluids are combined together in one, it is to be supposed that the particles of the same kind repel each other, and therefore take their stations accordingly.

Enough has been given to shew the method; it will be quite unnecessary to devise characters and combinations of them to exhibit to view in this way all the subjects that come under investigation; nor is it necessary to insist upon the accuracy of all these compounds, both in number and weight; the principle will be entered into more particularly hereafter, as far as respects the individual results. It is not to be understood that all those articles marked as simple substances, are necessarily such by the theory; they are only necessarily of such weights. Soda and Potash, such as they are found in combination with acids, are 28 and 42 respectively in weight; but according to Mr Davy's very important discoveries, they are metallic oxides; the former then must be considered as composed of an atom of metal, 21, and one of oxygen, 7; and the latter of an atom of metal, 35, and one of oxygen, 7. Or, soda contains 75 per cent. metal and 25 oxygen; potash, 83·3 metal and 16·7 oxygen. It is particularly remarkable, that according to the above-mentioned gentleman's essay on the Decomposition and Composition of the fixed alkalies, in the *Philosophical Transactions* (a copy of which essay he has just favoured me with) it appears that "the largest quantity of oxygen indicated by these experiments was, for potash 17, and for soda, 26 parts in 100, and the smallest 13 and 19."

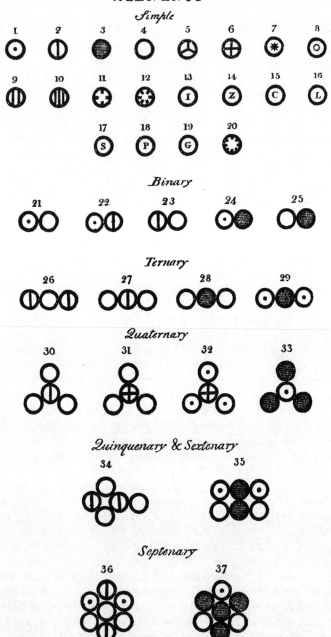

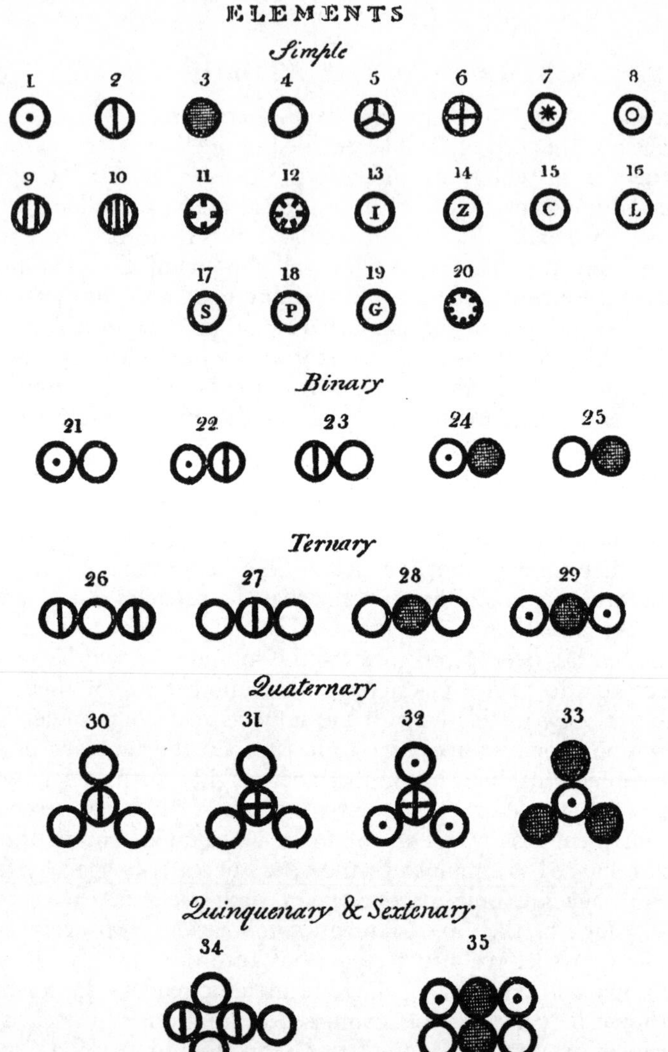

The figure [32] for sulphuretted hydrogen is incorrect: it ought to be 1 atom of hydrogen instead of 3, united to 1 of sulphur. See list.

The plate opposite contains the arbitrary marks or signs chosen to represent the several chemical elements or ultimate particles.

FIG.		FIG.	
1.	Hydrog. its rel. weight 1	11.	Strontites 46
2.	Azote 5	12.	Barytes...........................68
3.	Carbone or charcoal 5	13.	Iron 38
4.	Oxygen 7	14.	Zinc...............................56
5.	Phosphorus 9	15.	Copper............................56
6.	Sulphur 13	16.	Lead...............................96
7.	Magnesia 20	17.	Silver 100
8.	Lime 23	18.	Platina 100
9.	Soda 28	19.	Gold 140
10.	Potash 42	20.	Mercury 167

21\. An atom of water or steam, composed of 1 of oxygen and 1 of hydrogen, retained in physical contact by a strong affinity, and supposed to be surrounded by a common atmosphere of heat; its relative weight 8

22\. An atom of ammonia, composed of 1 of azote and 1 of hydrogen......6

23\. An atom of nitrous gas, composed of 1 of azote and 1 of oxygen...12

24\. An atom of olefiant gas, composed of 1 of carbone and 1 of hydrogen...6

25\. An atom of carbonic oxide composed of 1 of carbone and 1 of oxygen 12

26\. An atom of nitrous oxide, 2 azote + 1 oxygen...............................17

27\. An atom of nitric acid, 1 azote + 2 oxygen 19

28\. An atom of carbonic acid, 1 carbone + 2 oxygen 19

29\. An atom of carburetted hydrogen, 1 carbone + 2 hydrogen 7

30\. An atom of oxynitric acid, 1 azote + 3 oxygen 26

31\. An atom of sulphuric acid, 1 sulphur + 3 oxygen 34

32\. An atom of sulphuretted hydrogen, 1 sulphur + 1 hydrogen............16

33\. An atom of alcohol, 3 carbone + 1 hydrogen 16

34\. An atom of nitrous acid, 1 nitric acid + 1 nitrous gas 31

35\. An atom of acetous acid, 2 carbone + 2 water..............................26

36\. An atom of nitrate of ammonia, 1 nitric acid + 1 ammonia + 1 water 33

37\. An atom of sugar, 1 alcohol + 1 carbonic acid 35

THE COMBINATION OF GASES

Experiments on the relative weights in which chemical elements combine sufficed to lead Dalton to his modern revival of the atomic theory. But such experiments alone would have failed to bring to light all the laws of chemical action and all the inferences which may be drawn from them.

In a new problem it is always well to begin by examining the simplest case in which it appears. The great merit of the work of the distinguished French chemist Gay-Lussac on the combination of gases lies in his appreciation of the fact that in gases, or elastic fluids as he calls them, the disturbing force of cohesion is negligible, so that the phenomena of chemical combination appear in their simplest form.

MEMOIR ON THE COMBINATION OF GASEOUS SUBSTANCES WITH EACH OTHER

(*Read before the Philomathic Society*, 31*st Dec.*, 1808.)

By M. Gay-Lussac

Substances, whether in the solid, liquid, or gaseous state, possess properties which are independent of the force of cohesion; but they also possess others which appear to be modified by this force (so variable in its intensity), and which no longer follow any regular law. The same pressure applied to all solid or liquid substances would produce a diminution of volume differing in each case, while it would be equal for all elastic fluids. Similarly, heat expands all substances; but the dilatations of liquids and solids have hitherto presented no regularity, and it is only those of elastic fluids which are equal and independent of the nature of each gas. The attraction of the molecules in solids and liquids is, therefore, the cause which modifies their special properties; and it appears that it is only when the attraction is entirely destroyed, as in gases, that bodies under similar conditions obey simple and regular laws. At least, it is my intention to make known some new properties in gases, the effects of which are regular, by showing that these substances combine amongst themselves in very simple proportions, and that the contraction of volume which they experience on combination also follows a regular law. I hope by this means to give a proof of an idea advanced by several very distinguished chemists—that we are perhaps not far removed from the time when we shall be able to submit the bulk of chemical phenomena to calculation.

It is a very important question in itself, and one much discussed amongst chemists, to ascertain if compounds are formed in all sorts of proportions. M. Proust, who appears first to have fixed his attention on this subject, is of opinion that the metals are susceptible of only two degrees of oxidation, a *minimum* and a *maximum*; but led away by this seductive theory, he has seen himself forced to entertain principles contrary to physics in order to reduce to two oxides all those which the same metal sometimes presents. M. Berthollet thinks, on the other hand—reasoning from general considerations and his own experiments—that compounds are always formed in very variable proportions, unless they are determined by special causes, such as crystallisation, insolubility, or elasticity. Lastly, Dalton has advanced the idea that compounds of two bodies are formed in such a way that one atom of the one unites with one, two, three, or more atoms of the other. It would follow from this mode of looking at compounds that they are formed in constant proportions, the existence of intermediate bodies being excluded, and in this respect Dalton's theory would resemble that of M. Proust; but M. Berthollet has already strongly opposed it in the Introduction he has written to Thomson's *Chemistry*, and we shall see that in reality it is not entirely exact. Such is the state of the question now under discussion; it is still very far from receiving its solution, but I hope that the facts which I now proceed to set forth, facts which have entirely escaped the notice of chemists, will contribute to its elucidation.

Suspecting, from the exact ratio of 100 of oxygen to 200 of hydrogen, which M. Humboldt and I had determined for the proportions of water, that other gases might also combine in simple ratios, I have made the following experiments. I prepared fluoboric, muriatic, and carbonic gases, and made them combine successively with ammonia gas. 100 parts of muriatic gas saturate precisely 100 parts of ammonia gas, and the salt which is formed from them is perfectly neutral, whether one or the other of the gases is in excess. Fluoboric gas, on the contrary, unites in two proportions with ammonia gas. When the acid gas is put first into the graduated tube, and the other gas is then passed in, it is found that equal volumes of the two condense, and that the salt formed is neutral. But if we begin by first putting the ammonia gas into the tube, and then admitting the fluoboric gas in single bubbles, the first gas will then be in excess with regard to the second, and there will result a salt with excess of base, composed of 100 of fluoboric gas and 200 of ammonia gas. If carbonic gas is brought into contact with ammonia gas, by passing it sometimes first, sometimes second, into the tube, there is always formed a sub-carbonate composed of 100 parts of carbonic gas and 200 of ammonia gas. It may, however, be proved that neutral carbonate of ammonia would be composed of equal volumes of each of these components. M. Berthollet, who has analysed this salt, obtained by passing carbonic gas into the sub-carbonate, found that it was composed of 73·34 parts by weight of carbonic gas and 26·66 of ammonia gas. Now, if we suppose it to be composed of equal volumes of its components, we find from their known specific gravity, that it contains by weight

71·81	of carbonic acid,
28·19	of ammonia,
100·00	

a proportion differing only slightly from the preceding.

If the neutral carbonate of ammonia could be formed by the mixture of carbonic gas and ammonia gas, as much of one gas as of the other would be absorbed; and since we can only obtain it through the intervention of water, we must conclude that it is the affinity of this liquid which competes with that of the ammonia to overcome the elasticity of the carbonic acid, and that the neutral carbonate of ammonia can only exist through the medium of water.

Thus we may conclude that muriatic, fluoboric, and carbonic acids take exactly their own volume of ammonia gas to form neutral salts, and that the last two take twice as much to form *sub-salts*. It is very remarkable to see acids so different from one another neutralise a volume of ammonia gas equal to their own; and from this we may suspect that if all acids and all alkalis could be obtained in the gaseous state, neutrality would result from the combination of equal volumes of acid and alkali.

It is not less remarkable that, whether we obtain a neutral salt or a *sub-salt*, their elements combine in simple ratios which may be considered as limits to their proportions. Accordingly, if we accept the specific gravity of muriatic acid determined by M. Biot and myself, and those of carbonic gas and ammonia given by M. Biot and Arago, we find that dry muriate of ammonia is composed of

Ammonia,	100·0	or	38·35
Muriatic acid,	160·7		61·65
			100·00

a proportion very far from that of M. Berthollet—

100 of ammonia,
213 of acid.

In the same way, we find that sub-carbonate of ammonia contains

Ammonia,	100·0	or	43·98
Carbonic acid,	127·3		56·02
			100·00

and the neutral carbonate

Ammonia,	100·0	or	28·19
Carbonic acid,	254·6		71·81
			100·00

It is easy from the preceding results to ascertain the ratios of the capacity of fluoboric, muriatic, and carbonic acids; for since these three gases saturate the same volume of ammonia gas, their relative capacities will be inversely as their densities, allowance having been made for the water contained in muriatic acid.

We might even now conclude that gases combine with each other in very simple ratios; but I shall still give some fresh proofs.

According to the experiments of M. Amédée Berthollet, ammonia is composed of

100 of nitrogen
300 of hydrogen,

by volume.

I have found...that sulphuric acid is composed of

100 of sulphurous gas,
50 of oxygen gas.

When a mixture of 50 parts of oxygen and 100 of carbonic oxide (formed by the distillation of oxide of zinc with strongly calcined charcoal) is inflamed, these two gases are destroyed and their place taken by 100 parts of carbonic acid gas. Consequently carbonic acid may be considered as being composed of

100 of carbonic oxide gas,
50 of oxygen gas.

Davy, from the analysis of various compounds of nitrogen with oxygen, has found the following proportions by weight:

	Nitrogen	Oxygen
Nitrous oxide	63·30	36·70
Nitrous gas	44·05	55·95
Nitric acid	29·50	70·50

Reducing these proportions to volumes we find—

	Nitrogen	Oxygen
Nitrous oxide	100	49·5
Nitrous gas	100	108·9
Nitric acid	100	204·7

The first and the last of these proportions differ only slightly from 100 to 50, and 100 to 200; it is only the second which diverges somewhat from 100 to 100. The difference, however, is not very great, and is such as we might expect in experiments of this sort; and I have assured myself that it is actually nil. On burning the new combustible substance from potash in 100 parts by volume of nitrous gas, there remained over exactly 50 parts of nitrogen, the weight of which, deducted from that of the nitrous gas (determined with great care by M. Berard at Arcueil), yields as result that this gas is composed of equal parts by volume of nitrogen and oxygen.

We may then admit the following numbers for the proportions by volume of the compounds of nitrogen with oxygen:

	Nitrogen	Oxygen
Nitrous oxide	100	50
Nitrous gas	100	100
Nitric acid	100	200

...Thus it appears evident to me that gases always combine in the simplest proportions when they act on one another; and we have seen in reality in all the preceding examples that the ratio of combination is 1 to 1, 1 to 2, or 1 to 3. It is very important to observe that in considering weights there is no simple and finite relation between the elements of any one compound; it is only when there is a second compound between the same elements that the new proportion of the element that has been added is a multiple of the first quantity. Gases, on the contrary, in whatever proportions they may combine, always give rise to compounds whose elements by volume are multiples of each other.

ATOMS AND MOLECULES

Gay-Lussac's experiments on the combination of gases were explained in terms of Dalton's atomic theory by the Italian physicist Amedeo Avogadro (1776–1856) professor of physics at Turin.

He pointed out that the discovery of the simple ratios of volume in which gases combine leads at once to the view that equal volumes of all gases contain the same number of molecules. If we accept this view, say, for instance, in the case of oxygen and hydrogen, it follows from the volume of these gases and of the resultant compound gas (water vapour) that some molecules must consist of two half-molecules, which are the true units in chemical combination. Avogadro's half-molecules are Dalton's atoms, and the name of atom is now confined to them, the smallest parts of an element which can enter into a chemical reaction, while the name of molecule is reserved for the smallest part which can exist in the free state.

ESSAY ON A MANNER OF DETERMINING THE RELATIVE MASSES OF THE ELEMENTARY MOLECULES OF BODIES, AND THE PROPORTIONS IN WHICH THEY ENTER INTO THESE COMPOUNDS.

By A. Avogadro.

(*Journal de Physique*, 1811.)

I.

M. Gay-Lussac has shown in an interesting Memoir...that gases always unite in a very simple proportion by volume, and that when the result of the union is a gas, its volume also is very simply related to those of its components. But the quantitative proportions of substances in compounds seem only to depend on the relative number of molecules which combine, and on the number of composite molecules which result. It must then be admitted that very simple relations also exist between the volumes of gaseous substances and the numbers of simple or compound molecules which form them. The first hypothesis to present itself in this connection, and apparently even the only admissible one, is the supposition that the number of integral molecules in any gases is always the same for equal volumes, or always proportional to the volumes. Indeed, if we were to suppose that the number of molecules contained in a given volume were different for different gases, it would scarcely be possible to conceive that the law regulating the distance of molecules could give in all cases relations as simple as those which the facts just detailed compel us to acknowledge between the volume and the number of molecules....

Setting out from this hypothesis, it is apparent that we have the means of determining very easily the relative masses of the molecules of substances obtainable in the gaseous state, and the relative number of these molecules in compounds; for the ratios of the masses of the molecules are then the same as those of the densities of the different gases at equal temperature and pressure, and the relative number of molecules in a compound is given at once by the ratio of the volumes of the gases that form it. For example, since the numbers 1·10359 and 0·07321 express the densities of the two gases oxygen and hydrogen compared to that of atmospheric air as unity, and the ratio of the two numbers consequently represents the ratio between the masses of equal volumes of these two gases, it will also represent on our hypothesis the ratio of the masses of their molecules. Thus the mass of the molecule of oxygen will be about 15 times that of the molecule of hydrogen, or, more exactly, as 15·074 to 1. In the same way the mass of the molecule of nitrogen will be to that of hydrogen as 0·96913 to 0·07321, that is, as 13, or more exactly 13·238, to 1. On the other hand, since we know that the ratio of the volumes of hydrogen and oxygen in the formation of water is 2 to 1, it follows that water results from the union of each molecule of oxygen with two molecules of hydrogen. Similarly, according to the proportions by volume established by M. Gay-Lussac for the elements of ammonia, nitrous oxide, nitrous gas, and nitric acid, ammonia will result from the union of one molecule of nitrogen with three of hydrogen, nitrous oxide from one molecule of oxygen with two of nitrogen, nitrous gas from one molecule of nitrogen with one of oxygen, and nitric acid from one of nitrogen with two of oxygen.

II.

There is a consideration which appears at first sight to be opposed to the admission of our hypothesis with respect to compound substances. It seems that a molecule composed of two or more elementary molecules should have its mass equal to the sum of the masses of these molecules; and that in particular, if in a compound one molecule of one substance unites with two or more molecules of another substance, the number of compound molecules should remain the same as the number of molecules of the first substance. Accordingly, on our hypothesis, when a gas combines with two or more times its volume of another gas, the resulting compound, if gaseous, must have a volume equal to that of the first of these gases. Now, in general, this is not actually the case. For instance, the volume of water in the gaseous state is, as M. Gay-Lussac has shown, twice as great as the volume of oxygen which enters into it, or, what comes to the same thing, equal to that of the hydrogen instead of being equal to that of the oxygen. But a means of explaining facts of this type in conformity with our hypothesis presents itself naturally enough; we suppose, namely, that the constituent molecules of any simple gas whatever (i.e., the molecules which are at such a distance from each other that they cannot exercise their mutual action) are not formed of a solitary elementary molecule, but are made up of a certain number of these molecules united by attraction to form a single one; and further, that when molecules of another substance unite with the former to form a compound molecule, the integral molecule which should result splits up into two or more parts (or integral molecules) composed of half, quarter, &c., the number of elementary molecules going to form the constituent molecule of the first substance, combined with half, quarter, &c., the number of constituent molecules of the second substance that ought to enter into combination with one constituent molecule of the first substance (or, what comes to the same thing, combined with a number equal to this last of half-molecules, quarter-molecules, &c., of the second substance); so that the number of integral molecules of the compound becomes double, quadruple, &c., what it would have been if there had been no splitting up, and exactly what is necessary to satisfy the volume of the resulting gas.

Thus, for example, the integral molecule of water will be composed of a half-molecule of oxygen with one molecule, or, what is the same thing, two half-molecules of hydrogen.

On reviewing the various compound gases most generally known, I only find examples of duplication of the volume relatively to the volume of that one of the constituents which combines with one or more volumes of the other. We have already seen this for water. In the same way, we know that the volume of ammonia gas is twice that of the nitrogen which enters into it. M. Gay-Lussac has also shown that the volume of nitrous oxide is equal to that of the nitrogen which forms part of it, and consequently is twice that of the oxygen. Finally, nitrous gas, which contains equal volumes of nitrogen and oxygen, has a volume equal to the sum of the two constituent gases, that is to say, double that of each of them. Thus in all these cases there must be a division of the molecule into two; but it is possible that in other cases the division might be into four, eight, &c. The possibility of this division of compound molecules might have been conjectured *à priori*; for otherwise the integral molecules of bodies composed of several substances with a relatively large number of molecules, would come to have a mass excessive in comparison with the molecules of simple substances. We might therefore imagine that nature had some means of bringing them back to the order of the latter, and the facts have pointed out to us the existence of such means. Besides, there is another consideration which would seem to make us admit in some cases the division in question; for how could one otherwise conceive a real combination between two gaseous substances uniting in equal volumes without condensation, such as takes place in the formation of nitrous gas? Supposing the molecules to remain at such a distance that the mutual attraction of those of each gas could not be exercised, we cannot imagine that a new attraction could take place between the molecules of one gas and those of the other. But on the hypothesis of division of the molecule, it is easy to see that the combination really reduces two different molecules to one, and that there would be contraction by the whole volume of one of the gases if each compound molecule did not split up into two molecules of the same nature....

III.

Dalton, on arbitrary suppositions as to the most likely relative number of molecules in compounds, has endeavoured to fix ratios between the masses of the molecules of simple substances. Our hypothesis, supposing it well-founded, puts us in a position to confirm or rectify his results from precise data, and, above all, to assign the magnitude of compound molecules according to the volumes of the gaseous compounds, which depend partly on the division of molecules entirely unsuspected by this physicist.

Thus Dalton supposes that water is formed by the union of hydrogen and oxygen, molecule to molecule. From this, and from the ratio by weight of the two components, it would follow that the mass of the molecule of oxygen would be to that of hydrogen as $7\frac{1}{2}$ to 1 nearly, or, according to Dalton's evaluation, as 6 to 1. This ratio on our hypothesis is, as we saw, twice as great, namely, as 15 to 1. As for the molecule of water, its mass ought to be roughly expressed by $15 + 2 = 17$ (taking for unity that of hydrogen), if there were no division of the molecule into two; but on account of this division it is reduced to half, $8\frac{1}{2}$, or more exactly 8·537, as may also be found by dividing the density of aqueous vapour 0·625 (Gay-Lussac) by the density of hydrogen 0·0732. This mass only differs from 7, that assigned to it by Dalton, by the difference in the values for the composition of water; so that in this respect Dalton's result is approximately correct from the combination of two compensating errors,—the error in the mass of the molecule of oxygen, and his neglect of the division of the molecule.

Dalton supposes that in nitrous gas the combination of nitrogen and oxygen is molecule to molecule: we have seen on our hypothesis that this is actually the same. Thus Dalton would have found the same molecular mass for nitrogen as we have, always supposing that of hydrogen to be unity, if he had not set out from a different value for that of oxygen, and if he had taken precisely the same value for the quantities of the elements in nitrous gas by weight. But supposing the molecule of oxygen to be less than half what we find, he has been obliged to make that of nitrogen also equal to less than half the value we have assigned to it, viz., 5 instead of 13. As regards the molecule of nitrous gas itself, his neglect of the division of the molecule again makes his result approach ours; he has made it $6 + 5 = 11$, whilst according to us it is about $\frac{15 + 13}{2} = 14$, or more exactly $\frac{15{\cdot}974 + 13{\cdot}238}{2} = 14{\cdot}156$, as we also find by dividing 1·03636, the density of nitrous gas according to Gay-Lussac, by 0·07321. Dalton has likewise fixed in the same manner as the facts have given us, the relative number of molecules in nitrous oxide and in nitric acid, and in the first case the same circumstance has rectified his result for the magnitude of the molecule. He makes it $6 + 2 \times 5 = 16$, whilst according to our method it should be $\frac{15{\cdot}074 + 2 \times 13{\cdot}238}{2} = 20{\cdot}775$, a number which is also obtained by dividing 1·52092, Gay-Lussac's value for the density of nitrous oxide, by the density of hydrogen.

In the case of ammonia, Dalton's supposition as to the relative number of molecules in its composition is on our hypothesis entirely at fault. He supposes nitrogen and hydrogen to be united in it molecule to molecule, whereas we have seen that one molecule of nitrogen unites with three molecules of hydrogen. According to him the molecule of ammonia would be $5 + 1 = 6$; according to us it should be $\frac{13 + 3}{2} = 8$, or more exactly 8·119, as may also be deduced directly from the density of ammonia gas. The division of the molecule, which does not enter into Dalton's calculations, partly corrects in this case also the error which would result from his other suppositions.

All the compounds we have just discussed are produced by the union of one molecule of one of the components with one or more molecules of the other. In nitrous acid we have another compound of two of the substances already spoken of, in which the terms of the ratio between the number of molecules both differ from unity. From Gay-Lussac's experiments...it appears that this acid is formed from 1 part by volume of oxygen and 3 of nitrous gas, or, what comes to the same thing, of 3 parts of nitrogen and 5 of oxygen; hence it would follow, on our hypothesis, that its molecule should be composed of 3 molecules of nitrogen and 5 of oxygen, leaving the possibility of division out of account. But this mode of combination can be referred to the preceding simpler forms by considering it as the result of the union of 1 molecule of oxygen with 3 of nitrous gas, i.e. with 3 molecules, each composed of a half-molecule of oxygen and a half-molecule of nitrogen, which thus already includes the division of some of the molecules of oxygen which enter into that of nitrous acid. Supposing there to be no other division, the mass of this last molecule would be 57·542, that of hydrogen being taken as unity, and the density of nitrous acid gas would be 4·21267, the density of air being taken as unity. But it is probable that there is at least another division into two, and consequently a reduction of the density to half: we must wait until this density has been determined by experiment.

THE PERIODIC LAW

The theory of Avogadro enabled chemists to correct Dalton's atomic weights, and gradually, as new elements were discovered, a large amount of information on atomic weights accumulated. It became clear that the chemical properties of an element depended to some extent at any rate on its atomic weight, so that, if the elements were arranged in a table in order of ascending atomic weights, elements of similar chemical properties recurred at regular intervals in the table.

The first to point out clearly the great importance of this recurrence was the Russian chemist Mendeleeff, whose table was as below.

It will be seen that this arrangement brings elements of similar chemical properties, such as lithium, sodium and potassium, into the same vertical column. Such a periodicity of chemical properties naturally again calls to mind the idea that the various kinds of matter are composed of one primordial substance, the units of which are combined in schemes differing in the number and arrangement of the units.

TABLE. Distribution of the

Group	I		II		III		IV	
Series								
1	Hydrogen	1	—		—		—	
2	Lithium	7	Beryllium	9	Boron	11	Carbon	12
3	Sodium	23	Magnesium	24	Aluminium	27	Silicon	28
4	Potassium	39	Calcium	40	Scandium	44	Titanium	48
5	Copper	63	Zinc	65	Gallium	70	Germanium	72
6	Rubidium	85	Strontium	87	Yttrium	89	Zirconium	91
7	Silver	108	Cadmium	112	Indium	113	Tin	118
8	Caesium	133	Barium	137	Lanthanum	138	Cerium	130
9	—		—		—		—	
10	—		—		Ytterbium	173	—	
11	Gold	198	Mercury	200	Thallium	204	Lead	206
12	—		—		—		Thorium	232

THE PERIODIC LAW OF THE CHEMICAL ELEMENTS.

By D. MENDELEEFF.

(Faraday Lecture delivered before the Fellows of the Chemical Society in the theatre of the Royal Institution, on Tuesday, June 4, 1889: Appendix to English translation of Mendeleeff's Principles of Chemistry.)

...Before one of the oldest and most powerful of (scientific societies) I am about to take the liberty of passing in review the 20 years' life of a generalisation which is known under the name of the Periodic Law. It was in March 1869 that I ventured to lay before the then youthful Russian Chemical Society the ideas upon the same subject which I had expressed in my just written *Principles of Chemistry.*

Without entering into details, I will give the conclusions I then arrived at in the very words I used:

1. The elements, if arranged according to their atomic weights, exhibit an evident *periodicity* of properties.

2. Elements which are similar as regards their chemical pro-

LEMENTS IN GROUPS AND SERIES

V	VI	VII	VIII		
—	—	—	—		
Nitrogen 14	Oxygen 16	Fluorine 19			
Phosphorus 31	Sulphur 32	Chlorine 355			
Vanadium 51	Chromium 52	Manganese 55	Iron 56	Cobalt 58·5	Nickel 59
Arsenic 75	Selenium 79	Bromine 80			
Niobium 94	Molybdenum 96	—	Ruthenium 103	Rhodium 104	Palladium 106
Antimony 120	Tellurium 125	Iodine 127			
Didymium?	—	—	—	—	—
—	—	—			
Tantalum 182	Tungsten 185	—	Osmium 191	Iridium 193	Platinum 196
Bismuth 208	—	—			
—	Uranium 240	—			

perties have atomic weights which are either of nearly the same value (e.g. platinum, iridium, osmium) or which increase regularly (e.g. potassium, rubidium, caesium).

3. The arrangement of the elements, or of groups of elements, in the order of their atomic weights, corresponds to their so-called *valencies* as well as, to some extent, to their distinctive chemical properties—as is apparent, among other series, in that of lithium, beryllium, barium, carbon, nitrogen, oxygen, and iron.

4. The elements which are the most widely diffused have small atomic weights.

5. The *magnitude* of the atomic weight determines the character of the element, just as the magnitude of the molecule determines the character of a compound.

6. We must expect the discovery of many yet *unknown* elements—for example, elements analogous to aluminium and silicon, whose atomic weight would be between 65 and 75.

7. The atomic weight of an element may sometimes be amended by a knowledge of those of the contiguous elements. Thus, the atomic weight of tellurium must lie between 123 and 126, and cannot be 128.

8. Certain characteristic properties of the elements can be foretold from their atomic weights.

The aim of this communication will be fully attained if I succeed in drawing the attention of investigators to those relations which exist between the atomic weights of dissimilar elements, which, as far as I know, have hitherto been almost completely neglected. I believe that the solution of some of the most important problems of our science lies in researches of this kind.

To-day, twenty years after the above conclusions were formulated, they may still be considered as expressing the essence of the now well-known periodic law.

Reverting to the epoch terminating with the sixties, it is proper to indicate three series of data without the knowedge of which the periodic law could not have been discovered, and which rendered its appearance natural and intelligible.

In the first place, it was at that time that the numerical value of atomic weights became definitely known. Ten years earlier such knowledge did not exist, as may be gathered from the fact that in 1860 chemists from all parts of the world met at Karlsruhe in order to come to some agreement, if not with respect to views relating to atoms, at any rate as regards their definite representation. Many of those present probably remember how vain were the hopes of coming to an understanding, and how much ground was gained at that congress by the followers of the unitary theory so brilliantly represented by Cannizaro. I vividly remember the impression produced by his speeches, which admitted of no compromise, and seemed to advocate truth itself, based on the conceptions of Avogadro, Gerhardt, and Regnault, which at that time were far from being generally recognised. And though no understanding could be arrived at, yet the objects of the meeting were attained, for the ideas of Cannizaro proved, after a few years, to be the only ones which could stand criticism, and which represented an atom as—"the smallest portion of an element which enters into a molecule of its compound." Only such real atomic weights—not conventional ones—could afford a basis for generalisation. It is sufficient, by way of example, to indicate the following cases in which the relation is seen at once and is perfectly clear:

K = 39	Rb = 85	Cs = 133
Ca = 40	Sr = 87	Ba = 137

whereas with the equivalents then in use—

K = 39	Rb = 85	Cs = 133
Ca = 20	Sr = 43·5	Ba = 68·5

the consecutiveness of change in atomic weights, which with the true values is so evident, completely disappears.

Secondly, it had become evident during the period 1860–70, and even during the preceding decade, that the relations between the atomic weights of analogous elements were governed by some general and simple laws. Cooke, Cremers, Gladstone, Gmelin, Lenssen, Pettenkofer, and especially Dumas, had already established many facts bearing on that view. Thus Dumas compared the following groups of analogous elements with organic radicles:

	Diff.		Diff.		Diff.		Diff.
		Mg = 12		P = 31		O = 8	
			8		44		8
Li = 7		Ca = 20		As = 75		S = 16	
	16		3 × 8		44		3 × 8
Na = 23		Sr = 44		Sb = 119		Se = 40	
	16		3 × 8		2 × 44		3 × 8
K = 39		Ba = 68		Bi = 207		Te = 64	

and pointed out some really striking relationships, such as the following:

$$\begin{aligned} F &= 19 \\ Cl &= 35{\cdot}5 = 19 + 16{\cdot}5 \\ Br &= 80 = 19 + (2 \times 16{\cdot}5) + 28 \\ I &= 127 = (2 \times 19) + (2 \times 16{\cdot}5) + (2 \times 28). \end{aligned}$$

A. Strecker, in his work *Theorien und Experimente zur Bestimmung der Atomgewichte der Elemente* (Braunschweig, 1859), after summarising the data relating to the subject, and pointing out the remarkable series of equivalents—

$$Cr = 26{\cdot}2 \quad Mn = 27{\cdot}6 \quad Fe = 28 \quad Ni = 29 \quad Co = 30$$
$$Cu = 31{\cdot}7 \quad Zn = 32{\cdot}5$$

remarks that: "It is hardly probable that all the above-mentioned relations between the atomic weights (or equivalents) of chemically analogous elements are merely accidental. We must, however, leave to the future the discovery of the *law* of the relations which appear in these figures."

In such attempts at arrangement and in such views are to be recognised the real forerunners of the periodic law; the ground was prepared for it between 1860 and 1870, and that it was not expressed in a determinate form before the end of the decade may, I suppose, be ascribed to the fact that only analogous elements had been compared. The idea of seeking for a relation between the atomic weights of all elements was foreign to the ideas then current, so that neither the *vis tellurique* of De Chancourtois, nor the *law of octaves* of Newlands, could secure anybody's attention. And yet both De Chancourtois and Newlands, like Dumas and Strecker, more than Lenssen and Pettenkofer, had made an approach to the periodic law and had discovered its germs. The solution of the problem advanced but slowly, because the facts, and not the law, stood foremost in all attempts; and the law could not awaken a general interest so long as elements, having no apparent connection with each other, were included in the same octave, as for example:

1st octave of Newlands...	H	F	Cl	Co and Ni	Br	Pd	I	Pt & Ir
7th Ditto	O	S	Fe	Se	Rh & Ru	Te	Au	Os or Th

Analogies of the above order seemed quite accidental, and the more so as the octave contained occasionally ten elements instead of eight, and when two such elements as Ba and V, Co and Ni, or Rh and Ru, occupied one place in the octave. Nevertheless, the fruit was ripening, and I now see clearly that Strecker, De Chancourtois, and Newlands stood foremost in the way towards the discovery of the periodic law, and that they merely wanted the boldness necessary to place the whole question at such a height that its reflection on the facts could be clearly seen.

A third circumstance which revealed the periodicity of chemical elements was the accumulation, by the end of the sixties, of new information respecting the rare elements, disclosing their many-sided relations to the other elements and to each other. The researches of Marignac on niobium, and those of Roscoe on vanadium, were of special moment. The striking analogies between vanadium and phosphorus on the one hand, and between vanadium and chromium on the other, which became so apparent in the investigations connected with that element, naturally induced the comparison of V = 51 with Cr = 52, Nb = 94 with Mo = 96, and Ta = 192 with W = 194; while, on the other hand, P = 31 could be compared with S = 32, As = 75 with Se = 79, and Sb = 120 with Te = 125. From such approximations there remained but one step to the discovery of the law of periodicity.

The law of periodicity was thus a direct outcome of the stock of generalisations and established facts which had accumulated by the end of the decade 1860–1870: it is the embodiment of those data in a more or less systematic expression.

ELECTROCHEMISTRY

THE great advance in our knowledge of atomic theory which has distinguished recent years has been made by means of electrical evidence. We must therefore now turn back our attention to 1834, when Michael Faraday applied the new science of current electricity to chemical phenomena.

Faraday was assistant to Sir Humphry Davy, Professor of Chemistry at the Royal Institution, and succeeded his master in that Chair. Soon after Volta invented the galvanic battery in 1800, Davy, among the first, turned it to account. He decomposed the alkalies soda and potash, and extracted from them the new metals sodium and potassium. Probably it was these striking results which afterwards led Faraday to his researches on this subject.

At first, the ideas of static electricity and magnetism were used in examining the effects of currents. The metallic plates, by which the voltaic current enters and leaves a solution of a salt or acid, were regarded as centres of force analogous to the poles of a magnet, and it was Faraday who showed the advantage of a good nomenclature by inventing a new terminology and using it to develop new conceptions.

ON ELECTROCHEMICAL DECOMPOSITION

By MICHAEL FARADAY

(*Philosophical Transactions of the Royal Society*, 1834.)

PRELIMINARY

THE theory which I believe to be a true expression of the facts of electrochemical decomposition, and which I have therefore detailed in a former series of these Researches, is so much at variance with those previously advanced that I find the greatest difficulty in stating results, as I think, correctly, whilst limited to the use of terms which are current with a certain accepted meaning. Of this kind is the term *pole*, with its prefixes of positive and negative, and the attached ideas of attraction and repulsion. The general phraseology is that the positive pole *attracts* oxygen, acids, etc., or more cautiously, that it *determines* their evolution upon its surface; and that the negative pole acts in an equal manner upon hydrogen, combustibles, metals, and bases. According to my view, the determining force is *not* at the poles, but *within* the body under decomposition; and the oxygen and acids are rendered at the *negative* extremity of that body, whilst hydrogen, metals, etc., are evolved at the *positive* extremity.

To avoid, therefore, confusion and circumlocution, and for the sake of greater precision of expression than I can otherwise obtain, I have deliberately considered the subject with two friends, and with their assistance and concurrence in framing them, I purpose henceforward using certain other terms, which I will now define. The *poles*, as they are usually called, are only the doors or ways by which the electric current passes into and out of the decomposing body; and they of course, when in contact with that body, are the limits of its extent in the direction of the current. The term has been generally applied to the metal surfaces in contact with the decomposing substance; but whether philosophers generally would also apply it to the surfaces of air and water, against which I have effected electrochemical decomposition, is subject to doubt. In place of the term pole, I propose using that of *electrode**, and I mean thereby that substance, or rather surface, whether of air, water, metal, or any other body, which bounds the extent of the decomposing matter in the direction of the electric current.

The surfaces at which, according to common phraseology, the electric current enters and leaves a decomposing body are most important places of action, and require to be distinguished apart from the poles, with which they are mostly, and the electrodes, with which they are always, in contact.... The *anode*† is therefore that surface at which the electric current, according to our present expression, enters: it is the *negative* extremity of the decomposing body; is where oxygen, chlorine, acids, etc., are evolved; and is against or opposite the positive electrode. The *cathode*‡ is that surface at which the current leaves the decomposing body, and is its *positive* extremity; the combustible bodies, metals, alkalies, and bases are evolved there, and it is in contact with the negative electrode.

* *ἤλεκτρον*, and ὁδός, *a way*.
† *ἄνω*, *upwards*, and ὁδός, *a way*.
‡ *κατά*, *downwards*, and ὁδός, *a way*.

I shall have occasion in these Researches, also, to class bodies together according to certain relations derived from their electrical actions; and wishing to express those relations without at the same time involving the expression of any hypothetical views, I intend using the following names and terms. Many bodies are decomposed directly by the electric current, their elements being set free; these I propose to call *electrolytes* *....

Finally, I require a term to express those bodies which can pass to the *electrodes*, or, as they are usually called, the poles. Substances are frequently spoken of as being *electro-negative* or *electro-positive*, according as they go under the supposed influence of a direct attraction to the positive or negative pole. But these terms are much too significant for the use to which I should have to put them; for, though the meanings are perhaps right, they are only hypothetical, and may be wrong; and then, through a very imperceptible, but still very dangerous, because continual, influence, they do great injury to science by contracting and limiting the habitual views of those engaged in pursuing it. I propose to distinguish such bodies by calling those *anions*† which go to the *anode* of the decomposing body; and those passing to the *cathode*, *cations*‡; and when I have occasion to speak of these together, I shall call them *ions*. Thus, the chloride of lead is an *electrolyte*, and when *electrolyzed* evolves the two *ions*, chlorine and lead, the former being an *anion*, and the latter a *cation*.

These terms, being once well defined, will, I hope, in their use enable me to avoid much periphrasis and ambiguity of expression. I do not mean to press them into service more frequently than will be required, for I am fully aware that names are one thing and science another.

It will be well understood that I am giving no opinion respecting the nature of the electric current now, beyond what I have done on former occasions; and that though I speak of the current as proceeding from the parts which are positive to those which are negative, it is merely in accordance with the conventional, though in some degree tacit, agreement entered

* ἤλεκτρον, and λύω, *solvo*. Noun, electrolyte; verb, electrolyze.

† ἀνιών, *that which goes up*. [Neuter participle.]

‡ κατιών, *that which goes down*.

into by scientific men, that they may have a constant, certain, and definite means of referring to the direction of the forces of that current....

On a new measurer of volta-electricity

I have already said, when engaged in reducing common and voltaic electricity to one standard of measurement, and again when introducing my theory of electrochemical decomposition, that the chemical decomposing action of a current *is constant for a constant quantity of electricity*, notwithstanding the greatest variations in its sources, in its intensity, in the size of the *electrodes* used, in the nature of the conductors (or non-conductors) through which it is passed, or in other circumstances. The conclusive proofs of the truth of these statements shall be given almost immediately.

I endeavoured upon this law to construct an instrument which should measure out the electricity passing through it, and which, being interposed in the course of the current used in any particular experiment, should serve at pleasure, either as a *comparative standard* of effect or as a *positive measurer* of this subtile agent.

There is no substance better fitted, under ordinary circumstances, to be the indicating body in such an instrument as water; for it is decomposed with facility when rendered a better conductor by the addition of acids or salts: its elements may in numerous cases be obtained and collected without any embarrassment from secondary action, and, being gaseous, they are in the best physical condition for separation and measurement. Water, therefore, acidulated by sulphuric acid, is the substance I shall generally refer to, although it may become expedient in peculiar cases or forms of experiment to use other bodies.

The first precaution needful in the construction of the instrument was to avoid the recombination of the evolved gases, an effect which the positive electrode has been found so capable of producing. For this purpose various forms of decomposing apparatus were used. The first consisted of straight tubes, each containing a plate and wire of platina soldered together by gold, and fixed hermetically in the glass at the closed extremity of the tube (Fig. 1). The tubes were about 8 inches long, 0·7 of an inch in diameter, and graduated. The platina plates were about an inch long, as wide as the tubes would permit, and adjusted as near to the mouths of the tubes as was consistent with the safe collection of the gases evolved. In certain cases, where it was required to evolve the elements upon as small a surface as possible, the metallic extremity, instead of being a plate, consisted of the wire bent into the form of a ring (Fig. 2). When

Fig. 1.

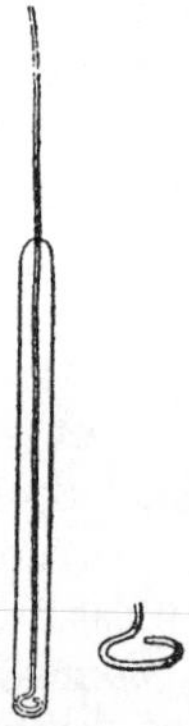

Fig. 2.

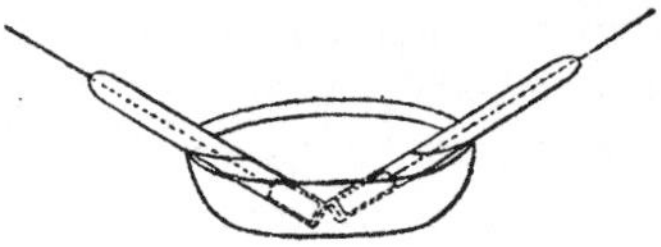

Fig. 3.

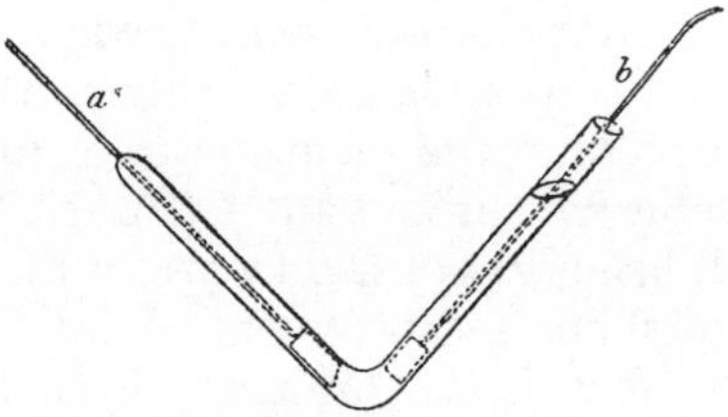

Fig. 4.

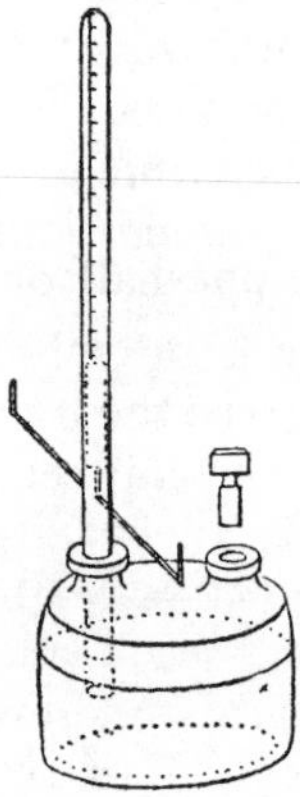

Fig. 5.

these tubes were used as measurers, they were filled with dilute sulphuric acid, inverted in a basin of the same liquid (Fig. 3), and placed in an inclined position, with their mouths near to each other, that as little decomposing matter should intervene as possible; and also, in such a direction that the platina plates should be in vertical planes.

Another form of apparatus is that delineated (Fig. 4). The tube is bent in the middle; one end is closed; in that end is fixed a wire and plate, *a*, proceeding so far downwards, that, when in the position figured, it shall be as near the angle as possible, consistently with the collection at the closed extremity of the tube, of all the gas evolved against it. The plane of this plate is also perpendicular. The other metallic termination, *b*, is introduced at the time decomposition is to be effected, being brought as near the angle as possible, without causing any gas to pass from it towards the closed end of the instrument. The gas evolved against it is allowed to escape.

The third form of apparatus contains both electrodes in the same tube; the transmission, therefore, of the electricity and the consequent decomposition is far more rapid than in the separate tubes. The resulting gas is the sum of the portions evolved at the two electrodes, and the instrument is better adapted than either of the former as a measurer of the quantity of voltaic electricity transmitted in ordinary cases. It consists of a straight tube (Fig. 5) closed at the upper extremity and graduated, through the sides of which pass platina wires (being fused into the glass), which are connected with two plates within. The tube is fitted by grinding into one mouth of a double-necked bottle. If the latter be one-half or two-thirds full of dilute sulphuric acid, it will, upon an inclination of the whole, flow into the tube and fill it. When an electric current is passed through the instrument, the gases evolved against the plates collect in the upper portion of the tube, and are not subject to the recombining power of the platina.... [Faraday then gives details of the experiments.]

I consider the foregoing investigation as sufficient to prove the very extraordinary and important principle with respect to WATER, *that when subjected to the influence of the electric current, a quantity of it is decomposed exactly proportionate to the quantity of electricity which has passed*, notwithstanding the thousand variations in the conditions and circumstances under which it may at the time be placed....

The instrument offers the only *actual measurer* of voltaic electricity which we at present possess. For without being at all affected by variations in time or intensity, or alterations in the current itself, of any kind, or from any cause, or even of intermissions of action, it takes note with accuracy of the quantity of electricity which has passed through it, and reveals that quantity by inspection; I have therefore named it a VOLTA-ELECTROMETER....

In the preceding cases, except the first, the water is believed to be inactive; but to avoid any ambiguity arising from its presence, I sought for substances from which it should be absent altogether; and taking advantage of the law of conduction* already developed, I soon found abundance, among which *protochloride of tin* was first subjected to decomposition in the following manner: A piece of platina wire had one extremity coiled up into a small knob, and having been carefully weighed, was sealed hermetically into a piece of bottle glass tube, so that the knob should be at the bottom of the tube within (Fig. 9). The tube was suspended by a piece of platina wire, so that the heat of a spirit-lamp could be applied to it. Recently fused protochloride of tin was introduced in sufficient quantity to occupy, when melted, about one-half of the tube; the wire of the tube was connected with a volta-electrometer, which was itself connected with the negative end of a voltaic battery: and a platina wire connected with the positive end of the same battery was dipped into the fused chloride in the tube; being, however, so bent that it could not by any shake of the hand or apparatus touch the negative electrode at the bottom of the vessel. The whole arrangement is delineated in Fig. 10.

Fig. 9.

Under these circumstances the chloride of tin was decomposed: the chlorine evolved at the positive electrode formed bichloride of tin, which passed away in fumes, and the tin evolved at the negative electrode combined with the platina, forming an alloy, fusible at the temperature to which the tube was subjected, and therefore never occasioning metallic communication through the decomposing chloride. When the experiment had been continued so long as to yield a reasonable quantity of gas in the volta-electrometer, the battery connection was broken, the positive electrode removed, and the tube and remaining chloride allowed to cool. When cold, the tube was broken open, the rest of the chloride and the glass being easily

* The law referred to asserts "the general assumption of conducting power by bodies as soon as they pass from the solid to the liquid state."

separable from the platina wire and its button of alloy. The latter when washed was then reweighed, and the increase gave the weight of the tin reduced.

I will give the particular results of one experiment, in illustration of the mode adopted in this and others, the results of which I shall have occasion to quote. The negative electrode weighed at first 20 grains; after the experiment it, with its button of alloy, weighed 23·2 grains. The tin evolved by the electric current at the *cathode* weighed therefore 3·2 grains. The quantity of oxygen and hydrogen collected in the volta-electrometer = 3·85 cubic inches. As 100 cubic inches of oxygen and hydrogen, in the proportions to form water, may be considered as weighing 12·92 grains, the 3·85 cubic inches

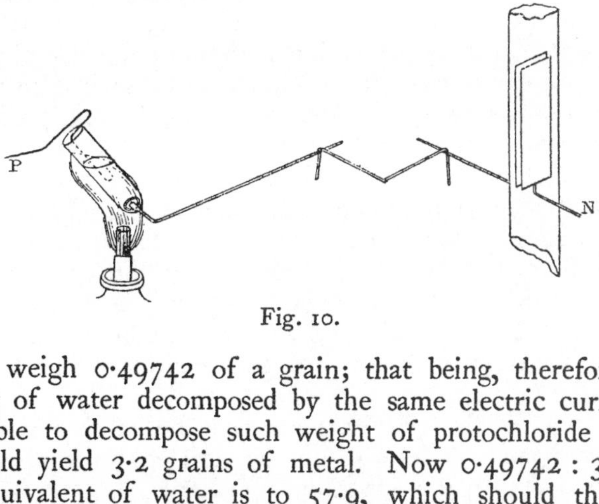

Fig. 10.

would weigh 0·49742 of a grain; that being, therefore, the weight of water decomposed by the same electric current as was able to decompose such weight of protochloride of tin as could yield 3·2 grains of metal. Now 0·49742 : 3·2 :: 9 the equivalent of water is to 57·9, which should therefore be the equivalent of tin, if the experiment had been made without error, and if the electrochemical decomposition *is in this case also definite*. In some chemical works 58 is given as the chemical equivalent of tin, in others 57·9. Both are so near to the result of the experiment, and the experiment itself is so subject to slight causes of variation (as from the absorption of gas in the volta-electrometer), that the numbers leave little doubt of the applicability of the *law of definite action* in this and all similar cases of electro-decomposition.

It is not often I have obtained an accordance in numbers so near as I have just quoted. Four experiments were made on the protochloride of tin; the quantities of gas evolved in the volta-electrometer being from 2·05 to 10·29 cubic inches. The average of the four experiments gave 58·43 as the electrochemical equivalent for tin.

The chloride remaining after the experiment was pure protochloride of tin; and no one can doubt for a moment that the equivalent of chlorine had been evolved at the *anode*, and, having formed bichloride of tin as a secondary result, had passed away.

Chloride of lead was experimented upon in a manner exactly similar, except that a change was made in the nature of the positive electrode; for as the chlorine evolved at the *anode* forms no perchloride of lead, but acts directly upon the platina, it produces, if that metal be used, a solution of chloride of platina in the chloride of lead; in consequence of which a portion of platina can pass to the *cathode*, and would then produce a vitiated result. I therefore sought for, and found in plumbago, another substance which could be used safely as the positive electrode in such bodies as chlorides, iodides, etc. The chlorine or iodine does not act upon it, but is evolved in the free state; and the plumbago has no reaction, under the circumstances, upon the fused chloride or iodide in which it is plunged. Even if a few particles of plumbago should separate by the heat or the mechanical action of the evolved gas, they can do no harm in the chloride.

The mean of three experiments gave the number of 100·85 as the equivalent of lead. The chemical equivalent is 103·5. The deficiency in my experiments I attribute to the solution of part of the gas in the volta-electrometer; but the results leave no doubt on my mind that both the lead and the chlorine are, in this case, evolved in *definite quantities* by the action of a given quantity of electricity.

THE IONIC DISSOCIATION THEORY

THE facts about electrolytic conduction, collected and put in order by Faraday and his successors, were carried a stage further by Hittorf, who pointed out that by measuring the unequal dilution of a solution round the two electrodes, the velocity relative to each other with which the ions moved could be calculated, and by Kohlrausch, who showed that, from these results and a measurement of the conductivity, the absolute velocity of the ions could be deduced. The idea of a certain freedom of migration was thus introduced, and in 1887 the facts and concepts of electrolysis were explained and connected with other physical and chemical phenomena by Arrhenius, a Swedish physicist. According to Arrhenius, the freedom of the opposite ions is not produced by the current, but exists by virtue of the properties of the electrolyte itself. The ions, to a large fraction of the whole number, are dissociated from each other, and the function of the current is merely to cause them to move in opposite directions.

If a solution be placed within a membrane permeable to the solvent but not to the solution, solvent will enter through the membrane till a certain equilibrium pressure, called osmotic pressure, is set up. This pressure is related to other properties of solutions, such as their vapour pressures and freezing points. Arrhenius shewed that each dissociated ion is not only an electric carrier, but is also chemically active, and moreover causes osmotic effects. Hence these apparently unrelated phenomena have been correlated. This ionic theory formed a ready-made explanation of the electric properties of gases when they came to be discovered.

ON THE DISSOCIATION OF SUBSTANCES DISSOLVED IN WATER

By SVANTE ARRHENIUS

Zeitschrift fur physikalische Chemie, I, 631, 1887. Translated by H. C. JONES: *Harper's Scientific Memoirs*, IV. New York and London, 1899.

IN a paper submitted to the Swedish Academy of Sciences, on the 14th of October, 1895, Van't Hoff proved experimentally, as well as theoretically, the following unusually significant generalization of Avogadro's law:

"The pressure which a gas exerts at a given temperature, if a definite number of molecules is contained in a definite volume, is equal to the osmotic pressure which is produced by most substances under the same conditions, if they are dissolved in any given liquid."

Van't Hoff has proved this law in a manner which scarcely leaves any doubt as to its absolute correctness. But a difficulty which still remains to be overcome, is that the law in question holds only for "most substances"; a very considerable number of the aqueous solutions investigated furnishing exceptions, and in the sense that they exert a much greater osmotic pressure than would be required from the law referred to.

If a gas shows such a deviation from the law of Avogadro, it is explained by assuming that the gas is in a state of dissociation. The conduct of chlorine, bromine, and iodine, at higher temperatures is a very well-known example. We regard these substances under such conditions as broken down into simple atoms.

The same expedient may, of course, be made use of to explain the exceptions to Van't Hoff's law; but it has not been put forward up to the present, probably on account of the newness of the subject, the many exceptions known, and the vigorous objections which would be raised from the chemical side to such an explanation. The purpose of the following lines is to show that such an assumption, of the dissociation of certain substances dissolved in water, is strongly supported by the conclusions drawn from the electrical properties of the same substances, and that also the objections to it from the chemical side are diminished on more careful examination.

In order to explain the electrical phenomena, we must assume with Clausius that some of the molecules of an electrolyte are dissociated into their ions, which move independently of one another....If, then, we could calculate what fraction of the molecules of an electrolyte is dissociated into ions, we should be able to calculate the osmotic pressure from Van't Hoff's law.

In a former communication "On the Electrical Conductivity of Electrolytes," I have designated those molecules whose ions are independent of one another in their movements, as active; the remaining molecules, whose ions are firmly combined with one another, as inactive. I have also maintained it as probable, that in extreme dilution all the inactive molecules of an electrolyte are transformed into active. This assumption I will make the basis of the calculations now to be carried out. I have designated the relation between the number of active molecules and the sum of the active and inactive molecules, as the activity coefficient. The activity coefficient of an electrolyte at infinite dilution is therefore taken as unity. For smaller dilution it is less than one, and, from the principles established in my work already cited, it can be regarded as equal to the ratio of the actual molecular conductivity of the solution to the maximum limiting value which the molecular conductivity of the same solution approaches with increasing dilution. This obtains for solutions which are not too concentrated (i.e., for solutions in which disturbing conditions, such as internal friction, etc., can be disregarded).

If this activity coefficient (α) is known, we can calculate as follows the value of the coefficient i tabulated by Van't Hoff. i is the relation between the osmotic pressure actually exerted by a substance and the osmotic pressure which it would exert if it consisted only of inactive (undissociated) molecules. i is evidently equal to the sum of the number of inactive molecules, plus the number of ions, divided by the sum of the inactive and active molecules. If, then, m represents the number of inactive, and n the number of active molecules, and k the number of ions into which every active molecule dissociates (e.g., $k = 2$ for KCl, i.e., K and Cl; $k = 3$ for $BaCl_2$ and K_2SO_4, i.e., Ba, Cl, Cl, and K, K, SO_4) then we have:

$$i = \frac{m + kn}{m + n}.$$

But since the activity coefficient α can evidently be written

$$\frac{n}{m + n},$$

we obtain $$i = 1 + (k - 1)\,\alpha.$$

Part of the figures given below (those in the last column) were calculated from this formula.

On the other hand, i can be calculated from the results of Raoult's experiments on the freezing points of solutions, making use of the principles stated by Van't Hoff. The lowering of the freezing point of water (in degrees Celsius) produced by dissolving a gram-molecule of the given substance in one litre of water, is divided by 18·5. The values of i thus calculated, are recorded in next to the last column. All the figures given below are calculated on the assumption that one gram of the substance to be investigated was dissolved in one litre of water as was done in the experiments of Raoult.

(90 substances are tabulated by Arrhenius, of which the following selection is made.)

Substance	Formula	a	$i=\frac{t}{18\cdot5}$	$i=1+(k-1)a$
Non-conductors				
Methyl alcohol	CH_3OH	0·00	0·94	1·00
Ethyl alcohol	C_2H_5OH	0·00	0·94	1·00
Glycerine	$C_3H_5(O_3H)$	0·00	0·92	1·00
Cane sugar	$C_{12}H_{22}O_{11}$	0·00	1·00	1·00
Phenol	C_6H_5OH	0·00	0·84	1·00
Electrolytes				
Sodium hydroxide	$NaOH$	0·88	1·96	1·88
Ammonia	NH_3	0·01	1·03	1·01
Hydrochloric acid	HCl	0·90	1·98	1·90
Sulphuric acid	H_2SO_4	0·60	2·06	2·19
Acetic acid	CH_3COOH	0·01	1·03	1·01
Potassium chloride	KCl	0·86	1·82	1·86
Sodium carbonate	Na_2CO_3	0·61	2·18	2·22
Copper sulphate	$CuSO_4$	0·35	0·97	1·35

An especially marked parallelism appears, beyond doubt, on comparing the figures in the last two columns. This shows, *a posteriori*, that in all probability the assumptions on which I have based the calculation of these figures are, in the main, correct. These assumptions were:

1. That Van't Hoff's law holds not only for most, but for all substances, even for those which have hitherto been regarded as exceptions (electrolytes in aqueous solution).

2. That every electrolyte (in aqueous solution), consists partly of active (in electrical and chemical relation), and partly of inactive molecules, the latter passing into active molecules

on increasing the dilution, so that in infinitely dilute solutions only active molecules exist.

The objections which can probably be raised from the chemical side are essentially the same which have been brought forward against the hypothesis of Clausius, and which I have earlier sought to show, were completely untenable. A repetition of these objections would, then, be almost superfluous. I will call attention to only one point. Although the dissolved substance exercises an osmotic pressure against the wall of the vessel, just as if it were partly dissociated into its ions, yet the dissociation with which we are here dealing is not exactly the same as that which exists when, e.g., an ammonium salt is decomposed at a higher temperature. The products of dissociation in the first case, the ions, are charged with very large quantities of electricity of opposite kind, whence certain conditions appear (the incompressibility of electricity), from which it follows that the ions cannot be separated from one another to any great extent without a large expenditure of energy. On the contrary, in ordinary dissociation where no such conditions exist, the products of dissociation can, in general, be separated from one another.

The above two assumptions are of the very widest significance, not only in their theoretical relation...but also, to the highest degree, in a practical sense. If it could, for instance, be shown that the law of Van't Hoff is generally applicable—which I have tried to show is highly probable—the chemist would have at his disposal an extraordinarily convenient means of determining the molecular weight of every substance soluble in a liquid.

THE ELECTRIC PROPERTIES OF GASES AND THE DISCOVERY OF PARTICLES SMALLER THAN CHEMICAL ATOMS

THE present era in physics began in 1895 with the discovery of X-rays by Professor Röntgen of Munich. When an electric discharge is passed through a highly exhausted glass bulb, the cathode or negative electrode becomes the source of rays, known as cathode rays, which fly from it in straight lines. If a metal plate, or anti-cathode, be placed opposite the cathode, it becomes the source of a new kind of radiation which was found by Röntgen to affect photographic plates through opaque screens. The influence of these rays on the electric properties of gases was examined systematically by Sir J. J. Thomson, now Master of Trinity College, Cambridge. He found that the passage of Röntgen rays through a rarefied gas made it a conductor of electricity, by forming charged ions from the molecules of the gas in some way then unknown.

Meanwhile, the phenomena and nature of cathode rays themselves had been for some time a subject of experiment and discussion. But in 1897 Thomson finally proved that they were flights of particles and surmised that they were more minute than atoms. It was soon shown that they possess a mass less (according to later measurements) than the eighteen hundredth part of the mass of a hydrogen atom. Thomson showed they were the same however produced and from whatever substance derived: they realise the vision of the ancient Greeks and the mediaeval alchemists of a common basis for matter.

CATHODE RAYS

By J. J. THOMSON, M.A., F.R.S.,
Professor of Physics, Cambridge.

Philosophical Magazine, Oct. 1897.

THE experiments discussed in this paper were undertaken in the hope of gaining some information as to the nature of the Cathode Rays. The most diverse opinions are held as to these rays; according to the almost unanimous opinion of German physicists they are due to some process in the æther to which—inasmuch as in a uniform magnetic field their course is circular and not rectilinear—no phenomenon hitherto observed is analogous: another view of these rays is that, so far from being wholly æthereal, they are in fact wholly material, and that they mark the paths of particles of matter charged with negative electricity. It would seem at first sight that it ought not to be difficult to discriminate between views so different, yet experience shows that this is not the case, as amongst the physicists who have most deeply studied the subject can be found supporters of either theory.

The electrified-particle theory has for purposes of research a great advantage over the ætherial theory, since it is definite and its consequences can be predicted; with the ætherial theory it is impossible to predict what will happen under any given circumstances, as on this theory we are dealing with hitherto unobserved phenomena in the æther, of whose laws we are ignorant.

The following experiments were made to test some of the consequences of the electrified-particle theory.

Charge carried by the Cathode Rays

If these rays are negatively electrified particles, then when they enter an enclosure they ought to carry into it a charge of negative electricity. This has been proved to be the case by Perrin, who placed in front of a plane cathode two coaxial metallic cylinders which were insulated from each other: the outer of these cylinders was connected with the earth, the inner with a gold-leaf electroscope. These cylinders were closed except for two small holes, one in each cylinder, placed so that the cathode rays could pass through them into the inside of the inner cylinder. Perrin found that when the rays passed into the inner cylinder the electroscope received a charge of negative electricity, while no charge went to the electroscope when the rays were deflected by a magnet so as no longer to pass through the hole.

This experiment proves that something charged with negative electricity is shot off from the cathode, travelling at right angles to it, and that this something is deflected by a magnet; it is open, however, to the objection that it does not prove that the cause of the electrification in the electroscope has anything to do with the cathode rays. Now the supporters of the ætherial theory do not deny that electrified particles are shot off from the cathode;

they deny, however, that these charged particles have any more to do with the cathode rays than a rifle-ball has with the flash when a rifle is fired. I have therefore repeated Perrin's experiment in a form which is not open to this objection. The arrangement used was as follows:—Two coaxial cylinders (Fig. 1) with slits in them are placed in a bulb connected with the discharge tube; the cathode rays from the cathode *A* pass into the bulb through a slit in a metal plug fitted into the neck of the tube; this plug is connected with the anode and is put to earth. The cathode rays thus do not fall upon the cylinders unless they are deflected by a magnet. The outer cylinder is connected with the earth, the inner with the electrometer. When the cathode rays (whose path was traced by the phosphorescence on the glass) did not fall on the slit, the electrical charge sent to the electrometer when the induction-coil producing the rays was set in action was small and irregular; when, however, the rays were bent by a magnet so as to fall on the slit there was a large charge of negative electricity sent to the electrometer. I was surprised at the magnitude of the charge; on some occasions enough negative electricity went through the narrow slit into the inner cylinder in one second to alter the potential of a capacity of 1·5 microfarads by 20 volts. If the rays were so much bent by the magnet that they overshot the slits in the cylinder, the charge passing into the cylinder fell again to a very small fraction of its value when the aim was true. Thus this experiment shows that however we twist and deflect the cathode rays by magnetic forces, the negative electrification follows the same path as the rays, and that this negative electrification is indissolubly connected with the cathode rays.

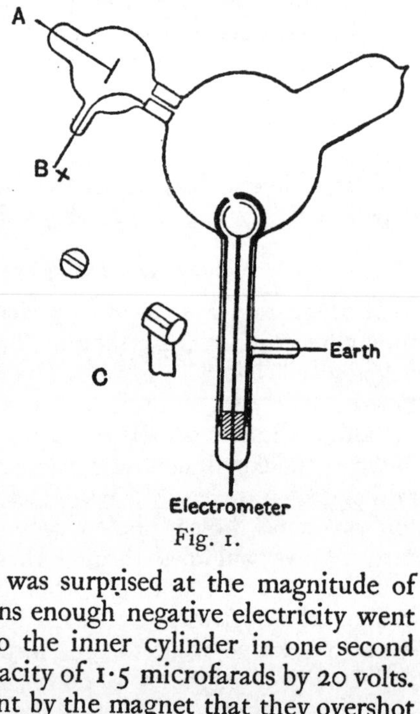

Fig. 1.

When the rays are turned by the magnet so as to pass through the slit into the inner cylinder, the deflexion of the electrometer connected with this cylinder increases up to a certain value, and then remains stationary although the rays continue to pour into the cylinder. This is due to the fact that the gas in the bulb becomes a conductor of electricity when the cathode rays pass through it, and thus, though the inner cylinder is perfectly insulated when the rays are not passing, yet as soon as the rays pass through the bulb the air between the inner cylinder and the outer one becomes a conductor, and the electricity escapes from the inner cylinder to the earth. Thus the charge within the inner cylinder does not go on continually increasing; the cylinder settles down into a state of equilibrium in which the rate at which it gains negative electricity from the rays is equal to the rate at which it loses it by conduction through the air. If the inner cylinder has initially a positive charge it rapidly loses that charge and acquires a negative one; while if the initial charge is a negative one, the cylinder will leak if the initial negative potential is numerically greater than the equilibrium value.

Deflexion of the Cathode Rays by an Electrostatic Field

An objection very generally urged against the view that the cathode rays are negatively electrified particles, is that hitherto no deflexion of the rays has been observed under a small electrostatic force, and though the rays are deflected when they pass near electrodes connected with sources of large differences of potential, such as induction-coils or electrical machines, the deflexion in this case is regarded by the supporters of the ætherial theory as due to the discharge passing between the electrodes, and not primarily to the electrostatic field. Hertz made the rays travel between two parallel plates of metal placed inside the discharge-tube, but found that they were not deflected when the plates were connected with a battery of storage cells; on repeating this experiment I at first got the same result, but subsequent experiments showed that the absence of deflexion is due to the conductivity conferred on the rarefied gas by the cathode rays. On measuring this conductivity it was found that it diminished very rapidly as the exhaustion increased; it seemed then that on trying Hertz's experiment at very high exhaustions there might be a chance of detecting the deflexion of the cathode rays by an electrostatic force.

The apparatus used is represented in Fig. 2.

The rays from the cathode *C* pass through a slit in the anode *A*, which is a metal plug fitting tightly into the tube and connected with the earth; after passing through a second slit in another earth-connected metal plug *B*, they travel between two parallel aluminium plates about 5 cm. long by 2 broad and at a distance of 1·5 cm. apart; they then fall on the end of the tube and

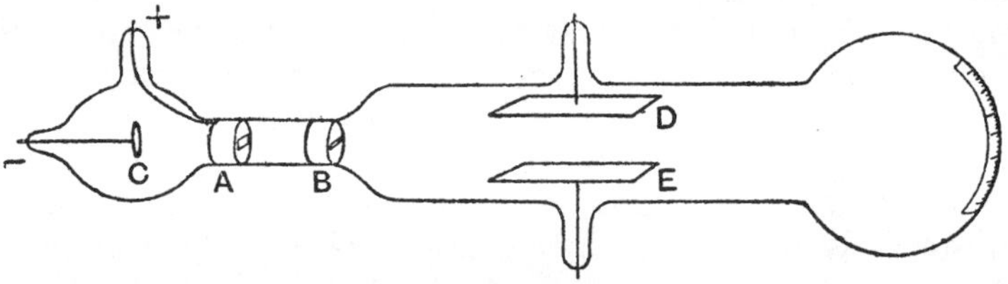

Fig. 2.

produce a narrow well-defined phosphorescent patch. A scale pasted on the outside of the tube serves to measure the deflexion of this patch. At high exhaustions the rays were deflected when the two aluminium plates were connected with the terminals of a battery of small storage-cells; the rays were depressed when the upper plate was connected with the positive, the lower with the negative pole. The deflexion was proportional to the difference of potential between the plates, and I could detect the deflexion when the potential-difference was as small as two volts....

Magnetic Deflexion of the Cathode Rays in Different Gases

The deflexion of the cathode rays by the magnetic field was studied with the aid of the apparatus shown in Fig. 4. The cathode was placed in a side-tube fastened on to a bell-jar; the opening between this tube and the bell-jar was closed by a metallic plug with a slit in it; this plug was connected with the earth and was used as the anode. The cathode rays passed through the slit in this plug into the bell-jar, passing in front of a vertical plate of glass ruled into small squares. The bell-jar was between two large parallel coils arranged as a Helmholtz galvanometer.

The course of the rays was determined by taking photographs of the bell-jar when the cathode rays were passing through it; the divisions on the plate enabled the path of the rays to be determined. Under the action of the magnetic field the narrow beam of cathode rays spreads out into a broad fan-shaped luminosity in the gas....One very interesting point brought out by the photographs is that in a given magnetic field, and with a given mean potential-difference between the terminals, the path of the rays is independent of the nature of the gas. Photographs were taken of the discharge in hydrogen, air, carbonic acid, methyl iodide, i.e., in gases whose densities range from 1 to 70, and yet, not only were the paths of the most deflected rays the same in all cases, but even the details, such as the distribution of the bright and dark spaces, were the same; in fact, the photographs could hardly be distinguished from each other. It is to be noted that the pressures were not the same; the pressures in the different gases were adjusted so that the mean potential-differences between the cathode and the anode were the same in all the gases. When the pressure of a gas is lowered, the potential-difference between the terminals increases, and the deflexion of the rays produced by a magnet diminishes, or at any rate the deflexion of the rays when the phosphorescence is a maximum diminishes. If an air-break is inserted an effect of the same kind is produced.

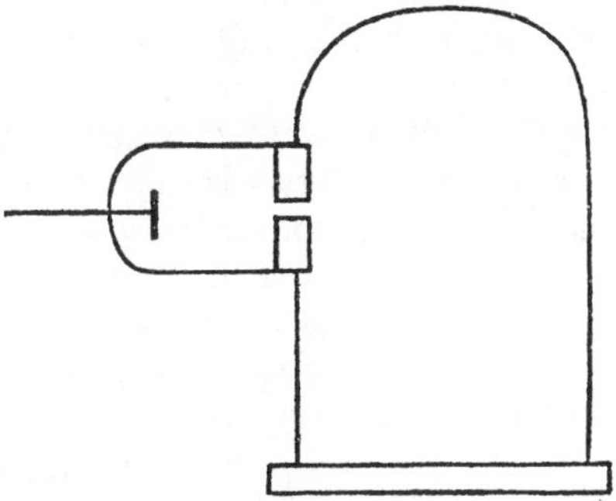

Fig. 4.

In the experiments with different gases, the pressures were as high as was consistent with the appearance of the phosphorescence on the glass, so as to ensure having as much as possible of the gas under consideration in the tube.

As the cathode rays carry a charge of negative electricity, are deflected by an electrostatic force as if they were negatively electrified, and are acted upon by a magnetic force in just the way in which this force would act on a negatively electrified body moving along the path of these rays, I can see no escape from the conclusion that they are charges of negative electricity carried by particles of matter. The question next arises, What are these particles? are they atoms, or molecules, or matter in a still finer state of subdivision? To throw some light on this point, I have made a series of measurements of the ratio of the mass of these particles to the charge carried by it. To determine this quantity, I have used two independent methods. The first of these is as follows:—Suppose we consider a bundle of homogeneous cathode rays. Let m be the mass of each of the particles, e the charge carried by it. Let N be the number of particles passing across any section of the beam in a given time; then Q the quantity of electricity carried by these particles is given by the equation $Ne = Q$. We can measure Q if we receive the cathode rays in the inside of a vessel connected with an electrometer. When these rays strike against a solid body, the temperature of the body is raised; the kinetic energy of the moving particles being converted into heat; if we suppose that all this energy is converted into heat, then if we measure the increase in the temperature of a body of known thermal capacity caused by the impact of these rays, we can determine W, the kinetic energy of the particles, and if v is the velocity of the particles,

$$\tfrac{1}{2}Nmv^2 = W.$$

If ρ is the radius of curvature of the path of these rays in a uniform magnetic field H, then

$$\frac{mv}{e} = H\rho = I,$$

where I is written for $H\rho$ for the sake of brevity. From these equations we get

$$\frac{1}{2}\frac{m}{e}v^2 = \frac{W}{Q},$$

$$v = \frac{2W}{QI},$$

$$\frac{m}{e} = \frac{I^2Q}{2W}.$$

Thus, if we know the values of Q, W, and I, we can deduce the values of v and m/e....

Before proceeding to discuss the results of these measurements I shall describe another method of measuring the quantities m/e and v of an entirely different kind from the preceding; this method is based upon the deflexion of the cathode rays in an electrostatic field. If we measure the deflexion experienced by the rays when traversing a given length under a uniform electric intensity, and the deflexion of the rays when they traverse a given distance under a uniform magnetic field, we can find the values of m/e and v in the following way:—

Let the space passed over by the rays under a uniform electric intensity F be l, the time taken for the rays to traverse this space is l/v, the velocity in the direction of F is therefore

$$\frac{Fe}{m}\frac{l}{v},$$

so that θ, the angle through which the rays are deflected when they leave the electric field and enter a region free from electric force, is given by the equation

$$\theta = \frac{Fe}{m}\frac{l}{v^2}.$$

If, instead of the electric intensity, the rays are acted on by a magnetic force H at right angles to the rays, and extending across the distance l, the velocity at right angles to the original path of the rays is

$$\frac{Hev}{m}\frac{l}{v},$$

so that ϕ, the angle through which the rays are deflected when they leave the magnetic field, is given by the equation

$$\phi = \frac{He}{m}\frac{l}{v}.$$

From these equations we get,

$$v = \frac{\phi}{\theta}\frac{F}{H},$$

and

$$\frac{m}{e} = \frac{H^2\theta\,.\,l}{F\phi^2}.$$

In the actual experiments H was adjusted so that $\phi = \theta$; in this case the equations become

$$v = \frac{F}{H},$$

$$\frac{m}{e} = \frac{H^2 l}{F\theta}.$$

The apparatus used to measure v and m/e by this means is that represented in Fig. 2. The electric field was produced by connecting the two aluminium plates to the terminals of a battery of storage cells. The phosphorescent patch at the end of the tube was deflected, and the deflexion measured by a scale pasted to the end of the tube. As it was necessary to darken the room to see the phosphorescent patch, a needle coated with luminous paint was placed so that by a screw it could be moved up and down the scale; this needle could be seen when the room was darkened, and it was moved until it coincided with the phosphorescent patch. Thus, when light was admitted, the deflexion of the phosphorescent patch could be measured.

The magnetic field was produced by placing outside the tube two coils whose diameter was equal to the length of the plates; the coils were placed so that they covered the space occupied by the plates, the distance between the coils was equal to the radius of either. The mean value of the magnetic force over the length l was determined in the following way: a narrow coil C whose length was l, connected with a ballistic galvanometer, was placed between the coils; the plane of the windings of C was parallel to the planes of the coils; the cross section of the coil was a rectangle 5 cm. by 1 cm. A given current was sent through the outer coils and the kick a of the galvanometer observed when this current was reversed. The coil C was then placed at the centre of two very large coils, so as to be in a field of uniform magnetic force: the current through the large coils was reversed and the kick b of the galvanometer again observed; by comparing a and b we can get the mean value of the magnetic force over a length l; this was found to be

$$60 \times i,$$

where i is the current flowing through the coils.

A series of experiments was made to see if the electrostatic deflexion was proportional to the electric intensity between the plates; this was found to be the case. In the following experiments the current through the coils was adjusted so that the electrostatic deflexion was the same as the magnetic.

Gas	θ	H	F	l	m/e	v
Air	8/110	5·5	$1{\cdot}5 \times 10^{10}$	5	$1{\cdot}3 \times 10^{-7}$	$2{\cdot}8 \times 10^{9}$
Air	9·5/110	5·4	$1{\cdot}5 \times 10^{10}$	5	$1{\cdot}1 \times 10^{-7}$	$2{\cdot}8 \times 10^{9}$
Air	13/110	6·6	$1{\cdot}5 \times 10^{10}$	5	$1{\cdot}2 \times 10^{-7}$	$2{\cdot}3 \times 10^{9}$
Hydrogen	9/110	6·3	$1{\cdot}5 \times 10^{10}$	5	$1{\cdot}5 \times 10^{-7}$	$2{\cdot}5 \times 10^{9}$
Carbonic acid...	11/110	6·9	$1{\cdot}5 \times 10^{10}$	5	$1{\cdot}5 \times 10^{-7}$	$2{\cdot}2 \times 10^{9}$
Air	6/110	5	$1{\cdot}8 \times 10^{10}$	5	$1{\cdot}3 \times 10^{-7}$	$2{\cdot}6 \times 10^{9}$
Air	7/110	3·6	1×10^{10}	5	$1{\cdot}1 \times 10^{-7}$	$2{\cdot}8 \times 10^{9}$

The cathode in the first five experiments was aluminium, in the last two experiments it was made of platinum; in the last experiment Sir William Crookes's method of getting rid of the mercury vapour by inserting tubes of pounded sulphur, sulphur iodide and copper filings between the bulb and the pump was adopted. In the calculation of m/e and v no allowance has been made for the magnetic force due to the coil in the region outside the plate; in this region the magnetic force will be in the opposite direction to that between the plates, and will tend to bend the cathode rays in the opposite direction: thus the effective value of H will be smaller than the value used in the equations, so that the values of m/e are larger, and those of v less than they would be if this correction were applied. This method of determining the values of m/e and v is much less laborious and probably more accurate than the former method; it cannot, however, be used over so wide a range of pressures.

From these determinations we see that the value of m/e is independent of the nature of the gas, and that its value 10^{-7} is very small compared with the value 10^{-4}, which is the smallest value of this quantity previously known, and which is the value for the hydrogen ion in electrolysis.

Thus for the carriers of electricity in the cathode rays m/e is very small compared with its value in electrolysis. The smallness of m/e may be due to the smallness of m or the largeness of e, or to a combination of these two. That the carriers of the charges in the cathode rays are small compared with ordinary molecules is shown, I think, by Lenard's results as to the rate at which the brightness of the phosphorescence produced by these rays diminishes with the length of path travelled by the ray. If we regard this phosphorescence as due to the impact of the charged particles, the distance through which the rays must travel before the phosphorescence fades to a given fraction (say $1/e$, where $e = 2{\cdot}71$) of its original intensity, will be some moderate multiple of the mean free path. Now Lenard found that this distance depends solely upon the density of the medium, and not upon its chemical nature or physical state. In air at atmospheric pressure the distance was about half a centimetre, and this must be comparable with the mean free path of the carriers through air at atmospheric pressure. But the mean free path of the molecules of air is a quantity of quite a different order. The carrier, then, must be small compared with ordinary molecules.

The two fundamental points about these carriers seem to me to be (1) that these carriers are the same whatever the gas through which the discharge passes, (2) that the mean free paths depend upon nothing but the density of the medium traversed by these rays.

It might be supposed that the independence of the mass of the carriers of the gas through which the discharge passes was due to the mass concerned being the quasi mass which a charged body possesses in virtue of the electric field set up in its neighbourhood; moving the body involves the production of a varying electric field, and, therefore, of a certain amount of energy which is proportional to the square of the velocity. This causes the charged body to behave as if its mass were increased by a quantity, which for a charged sphere is $\frac{1}{5}e^2/\mu a$, where e is the charge and a the radius of the sphere. If we assume that it is this mass which we are concerned with in the cathode rays, since m/e would vary as e/a, it affords no clue to the explanation of either of the properties (1 and 2) of these rays. This is not by any means the only objection to this hypothesis, which I only mention to show that it has not been overlooked.

The explanation which seems to me to account in the most

simple and straightforward manner for the facts is founded on a view of the constitution of the chemical elements which has been favourably entertained by many chemists: this view is that the atoms of the different chemical elements are different aggregations of atoms of the same kind. In the form in which this hypothesis was enunciated by Prout, the atoms of the different elements were hydrogen atoms; in this precise form the hypothesis is not tenable, but if we substitute for hydrogen some unknown primordial substance X, there is nothing known which is inconsistent with this hypothesis, which is one that has been recently supported by Sir Norman Lockyer for reasons derived from the study of the stellar spectra.

If, in the very intense electric field in the neighbourhood of the cathode, the molecules of the gas are dissociated and are split up, not into the ordinary chemical atoms, but into these primordial atoms, which we shall for brevity call corpuscles; and if these corpuscles are charged with electricity and projected from the cathode by the electric field, they would behave exactly like the cathode rays. They would evidently give a value of m/e which is independent of the nature of the gas and its pressure, for the carriers are the same whatever the gas may be; again, the mean free paths of these corpuscles would depend solely upon the density of the medium through which they pass. For the molecules of the medium are composed of a number of such corpuscles separated by considerable spaces; now the collision between a single corpuscle and the molecule will not be between the corpuscles and the molecule as a whole, but between this corpuscle and the individual corpuscles which make the molecule; thus the number of collisions the particle makes as it moves through a crowd of these molecules will be proportional, not to the number of the molecules in the crowd, but to the number of the individual corpuscles. The mean free path is inversely proportional to the number of collisions in unit volume; now as these corpuscles are all of the same mass, the number of corpuscles in unit volume will be proportional to the mass of unit volume, that is the mean free path will be inversely proportional to the density of the gas. We see, too, that so long as the distance between neighbouring corpuscles is large compared with the linear dimensions of a corpuscle, the mean free path will be independent of the way they are arranged, provided the number in unit volume remains constant, that is the mean free path will depend only on the density of the medium traversed by the corpuscles, and will be independent of its chemical nature and physical state: this from Lenard's very remarkable measurements of the absorption of the cathode rays by various media, must be a property possessed by the carriers of the charges in the cathode rays.

Thus on this view we have in the cathode rays matter in a new state, a state in which the subdivision of matter is carried very much further than in the ordinary gaseous state: a state in which all matter—that is, matter derived from different sources such as hydrogen, oxygen, &c.—is of one and the same kind; this matter being the substance from which all the chemical elements are built up.

POSITIVE RAYS AND ISOTOPES

Sir J. J. Thomson showed that negative or cathode rays were flights of the negative corpuscles, called by others electrons. He also investigated the nature of the positive rays which passed in the other direction in a vacuum tube. These positive rays are much less easily deflected, but by using very powerful electric and magnetic fields they have been examined and shown to consist of electrified atoms, the mass and charges of which may be estimated from the deflexions—a new and surprising method of chemical analysis.

In examining the gas neon, Thomson saw indications that it consisted of two constituents with atomic weights 20 and 22. Similar conclusions about other elements had already been reached from radio-active experiments. Soddy named such elements, similar in properties but different in atomic weight, isotopes.

Dr F. W. Aston, improving Thomson's method, has proved that many so-called elements really consist of mixtures of isotopes. The old puzzle of atomic weights not whole numbers thus finds its solution—they are average values between those of their constituents. All real elements have atomic weights that are very nearly indeed whole numbers if oxygen be taken as 16. For this work, Dr Aston was awarded the Nobel Prize for Chemistry in 1922, a distinction which had previously been given to Sir J. J. Thomson and Sir E. Rutherford.

ISOTOPES AND ATOMIC WEIGHTS

By Dr F. W. Aston

Nature: 105, p. 617.

In the atomic theory put forward by John Dalton in 1801 the second postulate was: "Atoms of the same element are similar to one another and equal in weight." For more than a century this was regarded by chemists and physicists alike as an article of scientific faith. The only item among the immense quantities of knowledge acquired during that productive period which offered the faintest suggestion against its validity was the inexplicable mixture of order and disorder among the elementary atomic weights. The general state of opinion at the end of last century may be gathered from the two following quotations from Sir William Ramsay's address to the British Association at Toronto in 1897:—

"There have been almost innumerable attempts to reduce the differences between atomic weights to regularity by contriving some formula which will express the numbers which represent the atomic weights with all their irregularities. Needless to say, such attempts have in no case been successful. Apparent success is always attained at the expense of accuracy, and the numbers reproduced are not those accepted as the true atomic weights. Such attempts, in my opinion, are futile. Still, the human mind does not rest contented in merely chronicling such an irregularity; it strives to understand why such an irregularity should exist....The idea...has been advanced by Prof. Schutzenburger, and later by Mr Crookes, that what we term the atomic weight of an element is a mean; that when we say the atomic weight of oxygen is 16, we merely state that the average atomic weight is 16; and it is not inconceivable that a certain number of molecules have a weight somewhat higher than 32, while a certain number have a lower weight."

This idea was placed on an altogether different footing some ten years later by the work of Sir Ernest Rutherford and his colleagues on radio-active transformations. The results of these led inevitably to the conclusion that there must exist elements which have chemical properties identical for all practical purposes, but the atoms of which have different weights. This conclusion has been recently confirmed in a most convincing manner by the production in quantity of specimens of lead from radio-active and other sources, which, though perfectly pure and chemically indistinguishable, give atomic weights differing by amounts quite outside the possible experimental error. Elements differing in mass but chemically identical and therefore occupying the same position in the periodic table have been called "isotopes" by Prof. Soddy.

At about the same period as the theory of isotopes was being developed by the radio-chemists at the heavy end of the periodic table an extremely interesting discovery was made by Sir J. J. Thomson, which carried the attack into the region of the lighter and non-radio-active elements. This was that, when positive rays from gases containing the element neon were analysed by electric and magnetic fields, results were obtained which indicated atomic weights roughly 20 and 22 respectively, the accepted atomic weight being 20·2. This naturally led to the expectation that neon might be a mixture of isotopes, but the weight 22 might possibly be due to other causes, and the method of analysis did not give sufficient accuracy to distinguish between 20 and 20·2 with certainty. Attempts were made to effect partial separation first by fractionation over charcoal cooled in liquid air, the results of which were absolutely negative, and then by diffusion, which in 1913 gave positive results, an apparent change in density of 0·7 per cent. between the lightest and heaviest fractions being attained after many thousands of operations. When the war interrupted the research, it might be said that several independent lines of reasoning pointed to the idea that neon was a mixture of isotopes, but that none of them could be said to carry the conviction necessary in such an important development.

By the time work was started again the isotope theory had been generally accepted so far as the radio-active elements were concerned, and a good deal of theoretical speculation had been made as to its applicability to the elements generally. As separation by diffusion is at the best extremely slow and laborious, attention was again turned to positive rays in the hope of increasing the accuracy of measurements to the required degree. This was done by means of the arrangement illustrated in Fig. 1.

Positive rays are sorted into an extremely thin ribbon by means of parallel slits S_1S_2, and are then spread into an electric spectrum by means of the charged plates P_1P_2. A portion of this spectrum deflected through an angle θ is selected by the diaphragm D and passed between the circular poles of a powerful electromagnet O the field of which is such as to bend the rays back again through an angle ϕ more than twice as great as θ. The result of this is that rays having a constant mass (or more correctly constant m/e) will converge to a focus F, and that if a

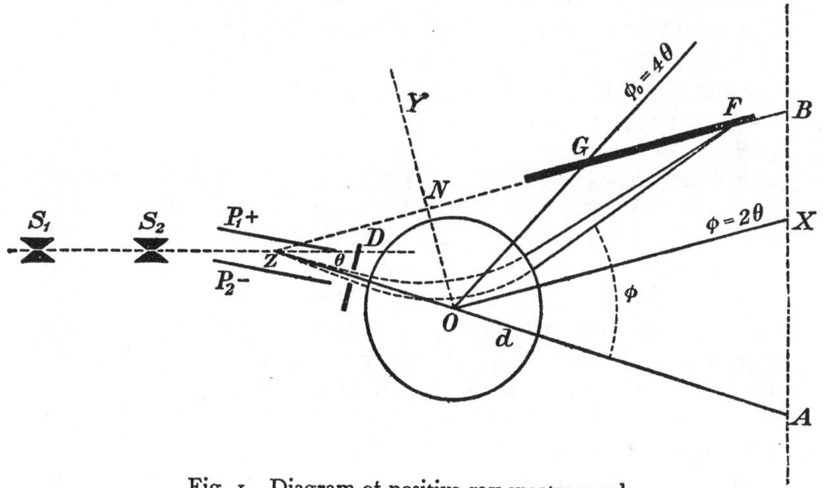

Fig. 1. Diagram of positive-ray spectrograph.

photographic plate is placed at GF as indicated, a *spectrum dependent on mass alone* will be obtained. On account of its analogy to optical apparatus, the instrument has been called a positive-ray spectrograph and the spectrum produced a mass-spectrum.

Plate IV shows a number of typical mass-spectra obtained by this means. The number above the lines indicates the masses they correspond to on the scale O = 16. It will be noticed that the displacement to the right with increasing mass is roughly linear. The measurements of mass made are not absolute, but relative to lines the mass of which is known. Such lines, due to hydrogen, carbon, oxygen, and their compounds, are generally present as impurities or purposely added, for pure gases are not suitable for the smooth working of the discharge tube. The two principal groups of these reference lines are the C_1 group due to C (12), CH (13), CH_2 (14), CH_3 (15), CH_4 or O (16), and the C_2 group 24–30 containing the very strong line 28, C_2H_4 or CO. In spectrum i. the presence of neon is indicated by the lines 20 and 22 situated between these groups. Comparative measurements show that these lines are 20·00, 22·00 with an accuracy of one-tenth per cent., which removes the last doubt as to the isotopic nature of neon.

The next element investigated was chlorine; this is characterised by four strong lines 35, 36, 37, 38, and fainter ones at 39, 40; there is no trace of a line at 35·46, the accepted atomic weight. From reasoning which cannot be given here in detail it seems certain that chlorine is a complex element, and consists of isotopes of atomic weights 35 and 37, with possibly another at 39. The lines at 36, 38 are due to the corresponding HCl's.

Particles with two, three, or more electronic charges will appear as though having half, a third, etc., their real mass. The corresponding lines are called lines of the second, third, or higher order. In spectrum ii. the lines of doubly charged chlorine atoms appear at 17·5 and 18·5. Analyses of argon indicate that this element consists almost entirely of atoms of weight 40, but a faint component 36 is also visible. Spectra v. and vi. are taken with this gas present; the former shows the interesting third order line at $13\frac{1}{2}$. Krypton and xenon give surprisingly complex results; the former is found to consist of no fewer than six isotopes, the latter of five (spectra viii. and ix.). Mercury is certainly a complex element probably composed of five or six isotopes, two of which have atomic weights 202 and 204; its multiply charged atoms give the imperfectly resolved groups, which are indicated in several of the spectra reproduced in Fig. 2.

By far the most important result obtained from this work is the generalisation that, with the exception of hydrogen, all the atomic weights of all elements so far measured are exactly whole numbers on the scale O = 16 to the accuracy of experiment (1 in 1000). By means of a special method (see *Phil. Mag.* May, 1920, p. 621), some results of which are given in spectrum vii., hydrogen is found to be 1·008, which agrees with the value accepted by chemists. This exception from the whole number rule is not unexpected, as on the Rutherford "nucleus" theory the hydrogen atom is the only one not containing any negative electricity in its nucleus.

The results which have so far been obtained with eighteen elements make it highly probable that the higher the atomic weight of an element, the more complex it is likely to be, and that there are more complex elements than simple. It must be noticed that, though the whole number rule asserts that a pure element must have a whole number atomic weight, there is no reason to suppose that all elements having atomic weights closely approximating to integers are therefore pure.

The very large number of different molecules possible when mixed elements combine to form compounds would appear to make their theoretical chemistry almost hopelessly complicated, but if, as seems likely, the separation of isotopes on any reasonable scale is to all intents impossible, their practical chemistry will not be affected, while the whole number rule introduces a very desirable simplification into the theoretical aspects of mass.

THE NATURE OF X-RAYS AND THE DISCOVERY OF ATOMIC NUMBERS

THE phenomena of light can only be explained by supposing that light is a wave motion. If a parallel beam of white light be allowed to fall on a grating, that is, a transparent or reflecting surface ruled with fine scratches of many thousands to the inch, the transmitted or reflected light will show the colours of the rainbow. From the angular deflexion of any particular colour its wave length can be measured in comparison with the distance between the scratches.

Röntgen rays did not show these diffraction spectra with ordinary gratings, and for some time their nature was doubtful. But it was suggested by Laue that the regular arrangement of atoms or molecules in a crystal might serve as a grating of very minute dimensions. X-rays have been examined in this way, especially by Sir Wm. Bragg and his son W. L. Bragg, and shown to be similar to waves of light of very short wave-length. The structure of crystals has also been elucidated.

The diffraction patterns of X-rays can be photographed, and hence their spectra examined. The spectra are found to depend on the nature of the anti-cathode which serves as the source of the rays, and, by using different elements as anti-cathode, their characteristic X-ray spectra can be mapped, the same crystal being employed as grating throughout.

A very interesting and important work on these lines was done by G. H. J. Moseley, of Eton and Oxford, who carried on his research in Sir E. Rutherford's laboratory at Manchester, joined the army in 1914 and lost in the war a life of the greatest promise to his country and the world.

Moseley discovered that the square roots of the frequency of vibration of the chief lines in the X-ray spectra increased regularly as he passed from element to element up the Periodic Table. He thus assigned to each element an atomic number, giving its true place in the Periodic Table, and the number of unit positive charges in the nucleus of its atom. This number is found to be fundamental in the modern theory of atomic structure and on it depend the chemical properties of the element.

The HIGH FREQUENCY SPECTRA OF THE ELEMENTS

By H. G. J. Moseley, M.A.

Phil. Mag. (1913), p. 1024.

In the absence of any available method of spectrum analysis, the characteristic types of X radiation, which an atom emits if suitably excited, have hitherto been described in terms of their absorption in aluminium. The interference phenomena exhibited by X-rays when scattered by a crystal have now, however, made possible the accurate determination of the frequencies of the various types of radiation. This was shown by W. H. and W. L. Bragg, who by this method analysed the line spectrum emitted by the platinum target of an X-ray tube. C. G. Darwin and the author extended this analysis and also examined the continuous spectrum, which in this case constitutes the greater part of the radiation. Recently Prof. Bragg has also determined the wave-lengths of the strongest lines in the spectra of nickel, tungsten, and rhodium. The electrical methods which have hitherto been employed are, however, only successful where a constant source of radiation is available. The present paper contains a description of a method of photographing these spectra, which makes the analysis of the X-rays as simple as any other branch of spectroscopy. The author intends first to make a general survey of the principal types of high-frequency radiation, and then to examine the spectra of a few elements in greater detail and with greater accuracy. The results already obtained show that such data have an important bearing on the question of the internal structure of the atom, and strongly support the views of Rutherford and of Bohr.

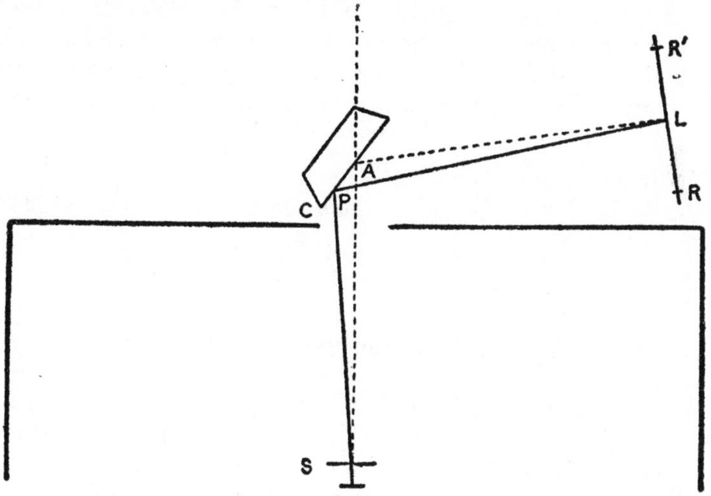

Fig. 1.

Kaye has shown that an element excited by a stream of sufficiently fast cathode rays emits its characteristic X radiation He used as targets a number of substances mounted on a truck inside an exhausted tube. A magnetic device enabled each target to be brought in turn into the line of fire. This apparatus was modified to suit the present work. The cathode stream was concentrated on to a small area of the target, and a platinum plate furnished with a fine vertical slit placed immediately in front of the part bombarded. The tube was exhausted by a Gaede mercury pump, charcoal in liquid air being also sometimes used to remove water vapour. The X-rays, after passing through the slit marked *S* in Fig. 1, emerged through an aluminium window ·02 mm. thick. The rest of the radiation was shut off by a lead box which surrounded the tube. The rays fell on the cleavage face, *C*, of a crystal of potassium ferrocyanide which was mounted on the prism-table of a spectrometer. The surface of the crystal was vertical and contained the geometrical axis of the spectrometer.

Now it is known that X-rays consist in general of two types, the heterogeneous radiation and characteristic radiations of definite frequency. The former of these is reflected from such a surface at all angles of incidence, but at the large angles used in the present work the reflexion is of very little intensity. The radiations of definite frequency, on the other hand, are reflected only when they strike the surface at definite angles, the glancing angle of incidence θ, the wave-length λ, and the "grating constant" d of the crystal being connected by the relation

$$n\lambda = 2d \sin \theta, \qquad \text{.................(1)}$$

where n, an integer, may be called the "order" in which the reflexion occurs. The particular crystal used, which was a fine specimen with face 6 cm. square, was known to give strong reflexions in the first three orders, the third order being the most prominent.

If then a radiation of definite wave-length happens to strike any part *P* of the crystal at a suitable angle, a small part of it is reflected. Assuming for the moment that the source of the radiation is a point, the locus of *P* is obviously the arc of a circle, and the reflected rays will travel along the generating lines of a cone with apex at the image of the source. The effect on a photographic plate *L* will take the form of the arc of an hyperbola, curving away from the direction of the direct beam. With a fine slit at *S*, the arc becomes a fine line which is slightly curved in the direction indicated.

The photographic plate was mounted on the spectrometer arm, and both the plate and the slit were 17 cm. from the axis. The importance of this arrangement lies in a geometrical property, for when these two distances are equal the point *L* at which a beam reflected at a definite angle strikes the plate is independent of the position of *P* on the crystal surface. The angle at which the crystal is set is then immaterial so long as a ray can strike some part of the surface at the required angle. The angle θ can be obtained from the relation $2\theta = 180° - SPL = 180° - SAL$.

The following method was used for measuring the angle *SAL*. Before taking a photograph a reference line *R* was made at both ends of the plate by replacing the crystal by a lead screen furnished with a fine slit which coincided with the axis of the spectrometer. A few seconds' exposure to the X-rays then gave a line *R* on the plate, and so defined on it the line joining *S* and *A*. A second line *RQ* was made in the same way after turning the spectrometer arm through a definite angle. The arm was then turned to the position required to catch the reflected beam and the angles *LAP* for any lines which were subsequently found on the plate deduced from the known value of *RAP* and the position of the lines on the plate. The angle *LAR* was measured with an error of not more than 0° . *D*, by superposing on the negative a plate on which reference lines had been marked in the same way at intervals of 1°. In finding from this the glancing angle of reflexion two small corrections were necessary in practice, since neither the face of the crystal nor the lead slit coincided accurately with the axis of the spectrometer. Wave-lengths varying over a range of about 30 per cent. could be reflected for a given position of the crystal.

In almost all cases the time of exposure was five minutes. Ilford X-ray plates were used and were developed with rodinal. The plates were mounted in a plate-holder, the front of which was covered with black paper. In order to determine the wave-length from the reflexion angle θ it is necessary to know both the order n in which the reflexion occurs and the grating constant d. n was determined by photographing every spectrum both in the second order and the third. This also gave a useful check on the accuracy of the measurements; d cannot be calculated directly for the complicated crystal potassium ferrocyanide. The grating constant of this particular crystal had, however, previously been accurately compared with d', the constant of a specimen of rock-salt. It was found that

$$d = 3d' \frac{\cdot 1988}{\cdot 1985}.$$

Now W. L. Bragg has shown that the atoms in a rock-salt

crystal are in simple cubical array. Hence the number of atoms per c.c.

$$2\frac{N\sigma}{M} = \frac{1}{(d')^3}.$$

N, the number of molecules in a gram-mol., $=6{\cdot}05\times 10^{23}$ assuming the charge on an electron to be $4{\cdot}89\times 10^{-10}$; σ, the density of this crystal of rock-salt, was $2{\cdot}167$, and M the molecular weight $=58{\cdot}46$.

This gives $d' = 2{\cdot}814\times 10^{-8}$ and $d = 8{\cdot}454\times 10^{-8}$ cm. It is seen that the determination of wave-length depends on $e^{\frac{1}{3}}$ so that the effect of uncertainty in the value of this quantity will not be serious. Lack of homogeneity in the crystal is a more likely source of error, as minute inclusions of water would make the density greater than that found experimentally.

Twelve elements have so far been examined....

Plate V shows the spectra in the third order placed approximately in register. Those parts of the photographs which represent the same angle of reflexion are in the same vertical line.... It is to be seen that the spectrum of each element consists of two lines. Of these the stronger has been called α in the table, and the weaker β. The lines found on any of the plates besides α and β were almost certainly all due to impurities. Thus in both the second and third order the cobalt spectrum shows Niα very strongly and Feα faintly. In the third order the nickel spectrum shows Mnα_2 faintly. The brass spectra naturally show α and β both of Cu and of Zn, but Znβ_2 has not yet been found. In the second order the ferro-vanadium and ferro-titanium spectra show very intense third-order Fe lines, and the former also shows Cuα_3 faintly. The Co contained Ni and 0·8 per cent. Fe, the Ni 2·2 per cent. Mn, and the V only a trace of Cu. No other lines have been found; but a search over a wide range of wave-lengths has been made only for one or two elements, and perhaps prolonged exposures, which have not yet been attempted, will show more complex spectra. The prevalence of lines due to impurities suggests that this may prove a powerful method of chemical analysis Its advantage over ordinary spectroscopic methods lies in the simplicity of the spectra and the impossibility of one substance masking the radiation from another. It may even lead to the discovery of missing elements, as it will be possible to predict the position of their characteristic lines....

Phil. Mag. (1914), p. 703.

The first part of this paper dealt with a method of photographing X-ray spectra, and included the spectra of a dozen elements. More than thirty other elements have now been investigated, and simple laws have been found which govern the results, and make it possible to predict with confidence the position of the principal lines in the spectrum of any element from aluminium to gold. The present contribution is a general preliminary survey, which claims neither to be complete nor very accurate.... .

The results obtained for radiations belonging to Barkla's K series are given in Table I, and for convenience the figures already given in Part I. are included. The wave-length λ has

Table I.

	α line $\lambda \times 10^8$ cm.	Q_K	N Atomic Number	β line $\lambda \times 10^8$ cm.
Aluminium	8·364	12·05	13	7·912
Silicon	7·142	13·04	14	6·729
Chlorine	4·750	16·00	17	—
Potassium	3·759	17·98	19	3·463
Calcium	3·368	19·00	20	3·094
Titanium	2·758	20·99	22	2·524
Vanadium	2·519	21·96	23	2·297
Chromium	2·301	22·98	24	2·093
Manganese	2·111	23·99	25	1·818
Iron	1·946	24·99	26	1·765
Cobalt	1·798	26·00	27	1·629
Nickel	1·662	27·04	28	1·506
Copper	1·549	28·01	29	1·402
Zinc	1·445	29·01	30	1·306
Yttrium	0·838	38·1	39	—
Zirconium	0·794	39·1	40	—
Niobium	0·750	40·2	41	—
Molybdenum	0·721	41·2	42	—
Ruthenium	0·638	43·6	44	—
Palladium	0·584	45·6	46	—
Silver	0·560	46·6	47	—

been calculated from the glancing angle of reflexion θ by means of the relation $n\lambda = 2 \sin \theta$, where d has been taken to be $8{\cdot}454 \times 10^{-8}$ cm. As before, the strongest line is called α and the next line β. The square root of the frequency of each line is plotted in Fig. 3, and the wave-lengths can be read off with the help of the scale at the top of the diagram.

The spectrum of Al was photographed in the first order only. The very light elements give several other fainter lines, which have not yet been fully investigated, while the results for Mg and Na are quite complicated, and apparently depart from the simple relations which connect the spectra of the other elements.

Table II

	α line $\lambda \times 10^8$ cm.	Q_L	N Atomic Number	β line $\lambda \times 10^8$ cm.	ϕ line $\lambda \times 10^8$ cm.	γ line $\lambda \times 10^8$ cm.
Zirconium	6·091	32·8	40	—	—	—
Niobium	5·749	33·8	41	5·507	—	—
Molybdenum	5·423	34·8	42	5·187	—	—
Ruthenium	4·861	36·7	44	4·660	—	—
Rhodium	4·622	37·7	45	—	—	—
Palladium	4·385	38·7	46	4·168	—	3·928
Silver	4·170	39·6	47	—	—	—
Tin	3·619	42·6	50	—	—	—
Antimony	3·458	43·6	51	3·245	—	—
Lanthanum	2·676	49·5	57	2·471	2·424	2·313
Cerium	2·567	50·6	58	2·366	2·315	2·209
Praseodymium	(2·471)	51·5	59	2·265	—	—
Neodymium	2·382	52·5	60	2·175	—	—
Samarium	2·208	54·5	62	2·008	1·972	1·893
Europium	2·130	55·5	63	1·925	1·888	1·814
Gadolinium	2·057	56·5	64	1·853	1·818	—
Holmium	1·914	58·6	66	1·711	—	—
Erbium	1·790	60·6	68	1·591	1·563	—
Tantalum	1·525	65·6	73	1·330	—	1·287
Tungsten	1·486	66·5	74	—	—	—
Osmium	1·397	68·5	76	1·201	—	1·172
Iridium	1·354	69·6	77	1·155	—	1·138
Platinum	1·316	70·6	78	1·121	—	1·104
Gold	1·287	71·4	79	1·092	—	1·078

In the spectra from yttrium onwards only the α line has so far been measured, and further results in these directions will be given in a later paper. The spectra both of K and of Cl were obtained by means of a target of KCl, but it is very improbable that the observed lines have been attributed to the wrong elements. The α line for elements from Y onwards appeared to consist of a very close doublet, an effect previously observed by Bragg in the case of rhodium.

The results obtained for the spectra of the L series are given in Table II and plotted in Fig. 3. These spectra contain five lines, α, β, γ, δ, ϵ, reckoned in order of decreasing wave-length and decreasing intensity. There is also always a faint companion α' on the long wave-length side of α, a rather faint line ϕ between β and γ for the rare earth elements at least, and a number of very faint lines of wave-length greater than α. Of these, α, β, ϕ, and γ have been systematically measured with the object of finding out how the spectrum alters from one element to another. The fact that often values are not given for all these lines merely indicates the incompleteness of the work. The spectra, so far as they have been examined, are so entirely similar that without doubt α, β, and γ at least always exist. Often γ was not included in the limited range of wave-lengths which can be photographed on one plate. Sometimes lines have not been measured, either on account of faintness or of the confusing proximity of lines due to impurities....

Conclusions.

In Fig. 3 the spectra of the elements are arranged on horizontal lines spaced at equal distances. The order chosen for the elements is the order of the atomic weights, except in the cases of A, Co, and Te, where this clashes with the order of the chemical properties. Vacant lines have been left for an element between Mo and Ru, an element between Nd and Sa, and an element between W and Os, none of which are yet known, while Tm, which Welsbach has separated into two constituents, is given two lines. This is equivalent to assigning to successive elements a series of successive characteristic integers. On this principle the integer N for Al, the thirteenth element, has been taken to be 13, and the values of N then assumed by the other elements

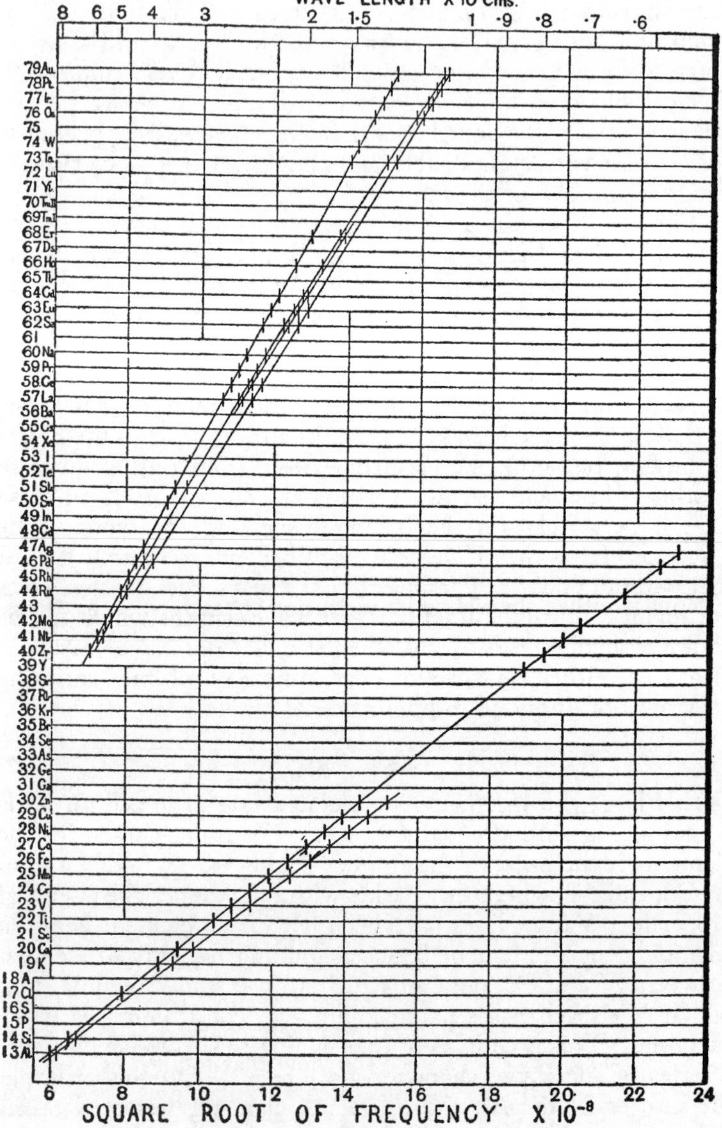
WAVE LENGTH X 10^8 Cms.
8 6 5 4 3 2 1·5 1 ·9 ·8 ·7 ·6
79 Au
78 Pt
77 Ir
76 Os
75
74 W
73 Ta
72 Lu
71 Yt
70 Tm II
69 Tm I
68 Er
67 Ds
66 Ho
65 Tb
64 Gd
63 Eu
62 Sa
61
60 Nd
59 Pr
58 Ce
57 La
56 Ba
55 Cs
54 Xe
53 I
52 Te
51 Sb
50 Sn
49 In
48 Cd
47 Ag
46 Pd
45 Rh
44 Ru
43
42 Mo
41 Nb
40 Zr
39 Y
38 Sr
37 Rb
36 Kr
35 Br
34 Se
33 As
32 Ge
31 Ga
30 Zn
29 Cu
28 Ni
27 Co
26 Fe
25 Mn
24 Cr
23 V
22 Ti
21 Sc
20 Ca
19 K
18 A
17 Cl
16 S
15 P
14 Si
13 Al
6 8 10 12 14 16 18 20 22 24
SQUARE ROOT OF FREQUENCY X 10^-8

Fig. 3.

are given on the left-hand side of Fig. 3. This proceeding is justified by the fact that it introduces perfect regularity into the X-ray spectra. Examination of Fig. 3 shows that the values of $\nu^{\frac{1}{2}}$ for all the lines examined both in the K and the L series now fall on regular curves which approximate to straight lines. The same thing is shown more clearly by comparing the values of N in Table I with those of

$$Q_K = \sqrt{\frac{\nu}{\frac{3}{4}\nu_0}},$$

ν being the frequency of the line and ν_0 the fundamental Rydberg frequency. It is here plain that $Q_K = N - 1$ very approximately, except for the radiations of very short wave-length which gradually diverge from this relation. Again, in Table II a comparison of N with

$$Q_L = \sqrt{\frac{\nu}{\frac{5}{36}\nu_0}},$$

where ν is the frequency of the La line, shows that $Q_L = N - 7{\cdot}4$ approximately, although a systematic deviation clearly shows that the relation is not accurately linear in this case.

Now if either the elements were not characterized by these integers, or any mistake had been made in the order chosen or in the number of places left for unknown elements, these regularities would at once disappear. We can therefore conclude from the evidence of the X-ray spectra alone, without using any theory of atomic structure, that these integers are really characteristic of the elements. Further, as it is improbable that two different stable elements should have the same integer, three, and only three, more elements are likely to exist between Al and Au. As the X-ray spectra of these elements can be confidently predicted, they should not be difficult to find. The examination of keltium would be of exceptional interest, as no place has been assigned to this element.

Now Rutherford has proved that the most important constituent of an atom is its central positively charged nucleus, and van den Broek has put forward the view that the charge carried by this nucleus is in all cases an integral multiple of the charge on the hydrogen nucleus. There is every reason to suppose that the integer which controls the X-ray spectrum is the same as the number of electrical units in the nucleus, and these experiments therefore give the strongest possible support to the hypothesis of van den Broek. Soddy has pointed out that the chemical properites of the radio-elements are strong evidence that this hypothesis is true for the elements from thallium to uranium, so that its general validity would now seem to be established.

RADIO-ACTIVITY AND THE STRUCTURE OF THE ATOM

RADIO-ACTIVITY was discovered by Becquerel, who found that uranium and its compounds gave out a hitherto unknown kind of radiation. M. and Madame Curie, by separating the most active part of certain uranium minerals in which this activity was present to a degree greater than in uranium, isolated the salt of a new and intensely active element which they called radium.

The explanation of the new phenomena was given by Rutherford and Soddy, who pointed out that radio-activity was always accompanied by the formation of new chemical substances, that it depended on the elements present and not on their state of combination, and that the amounts of energy evolved were immensely greater than those associated with any known chemical change. They concluded that radio-activity was due to the spontaneous explosion of atoms into atoms of less atomic weight, and particles shot forth with great velocity. Of these particles some, the so-called α-rays, are helium atoms, and others, the β rays, are identical in nature with the electrons of cathode rays. The line of research thus opened up has led to theories concerning the structure of the atom. Of these the most generally accepted is that chiefly due to Sir Ernest Rutherford, now Cavendish Professor of Physics at Cambridge.

X-rays, light, and the electromagnetic waves used in wireless telegraphy are the same in kind, differing only in the length of the waves and the frequency of the vibrations. The spectra of light and of X-rays are characteristic of the elements from which they arise. Hence they are atomic phenomena, and the parts of the atoms which send them forth must be electrical. We thus reach by another road, opened up by Lorentz and Larmor, the electron discovered by Thomson in cathode rays, and are led to an electrical theory of matter.

Thomson's electron is a negative unit of electricity, and, if it is to be imagined as part of a neutral atom, that atom must contain equivalent positive electricity. The deflexion or scattering of α particles from radio-active substances by collision with atoms indicates that this positive electricity exists as a very minute central nucleus, able to exert very intense forces on the α particles. By these and many other laborious experiments, the innermost structure of the atom has been investigated.

An α particle, flying with a speed approaching that of light, is the most intense concentration of energy known to us. If anything can smash an atom into fragments, it would be the impact of an α particle. And Rutherford has now found evidence that some atoms are thus broken up.

The dream of the mediaeval alchemist is realised. Not only do radio-active elements disintegrate spontaneously—that seems beyond human control—but, by a bombardment of α-rays, infinitely little, and yet infinitely more intense in its own minute field than an artillery or high explosive, man can produce these changes at will, and gain at last his age-long desire, the transmutation of the elements.

THE STABILITY OF ATOMS

By Sir Ernest Rutherford

(Abridged Report of a Lecture delivered before the Physical Society on June 10, 1921.)

During the latter half of the nineteenth century it was generally accepted that the atoms of the chemist and physicist were permanent and indestructible, and were uninfluenced by the most drastic physical and chemical agencies available. The existence of elements in the earth that appeared to have suffered no change within periods of time measured by the geological epochs gave a strong support to the prevailing view of the inherent stability of the elements. The discovery at the beginning of the twentieth century that the radio-active elements uranium and thorium were undergoing a veritable transformation, spontaneous and quite uncontrollable by the agencies at our disposal, was the first serious shock to our belief in the permanency of the elements. The essential phenomena which accompanied the series of transformations soon became clear. The disintegration of an atom was accompanied either by the emission of a swift atom of helium carrying a positive charge, or of a swift electron. With the exception possibly of potassium and rubidium, only the heavy radio-active elements showed this lack of stability. The great majority of the chemical elements appeared, as before, to be inherently stable and to be unaffected by the most intense forces at our disposal.

A number of attempts have been made from time to time to test whether the atoms of the elements can be broken up by artificial methods. Some thought they had obtained evidence of the production of hydrogen and helium in the electric discharge tube. It is, however, a matter of such great difficulty to prove the absence of these elements as a contamination in the materials used that the evidence of transformation has not carried conviction to the minds of the majority of scientific men.

In this lecture an account will be given of some preliminary experiments which indicate the possibility of artificial disintegration of some of the ordinary elements by a new method. Before discussing the results, it is desirable to say a few words on the modern conception of the structure of the atom. The results have been interpreted on the nuclear theory of atomic constitution. According to this view, the atom is to be regarded as consisting of a minute positively charged nucleus, in which most of the mass of the atom is concentrated, surrounded at a distance by a distribution of negative electrons which make the atom electrically neutral. We know that one or more of these outer electrons can be easily removed from the atom. The atom thus undergoes a kind of transformation, but only a temporary one, for the missing electrons are readily recaptured from neighbouring atoms. It seems not unlikely that the whole of the exterior electrons might be removed from an atom without interfering sensibly with the stability of its nucleus. Under suitable conditions, the atom would promptly regain its lost electrons and be indistinguishable from the original atom. In order to produce a *permanent* transformation of the atom, it is necessary to disintegrate the nucleus. Such a disruption of the nucleus occurs spontaneously in the radio-active atoms, and the processes appear to be irreversible under ordinary conditions.

The nucleus, however, is very small, and its constituent parts are probably held together by strong forces; and only a few agencies are available for an attack on its structure. The most concentrated source of energy at our command is a swift α-particle, and it is to be expected that an α-particle would occasionally approach so close to the nucleus as to disintegrate its structure. It is, indeed, from a study of the deflexion of swift α-particles in passing through matter that we have obtained the strongest evidence in support of the theory of the nuclear constitution of atoms. In the region surrounding a heavy nucleus, the inverse square law holds for the forces of repulsion between the charged α-particle and positively charged nucleus. The particle describes a hyperbolic orbit round the nucleus, and the amount of its deflection depends on the closeness of its approach. It is from a study of this scattering of α-particles, combined with Moseley's interpretation of the X-ray spectra of the elements, that we know the magnitude of the resultant positive charge on the nucleus. This charge, in fundamental units, is equal to the atomic or ordinal number of the element, and varies between 1 for hydrogen and 92 for uranium. Recently Chadwick has shewn by direct measurements of the scattering of α-particles that the charge on the nucleus is in close accord with Moseley's deduction, and has thus verified the correctness of this fundamental conclusion.

Some information about the dimensions of the nucleus can be obtained by studying the amount of scattering of α-particles at large angles by different atoms. The general results indicate that the nucleus of a heavy atom, if assumed spherical, cannot have a radius greater than 6×10^{-12} cm. It is not unlikely that the dimensions may be smaller than this. No doubt the size of a nucleus decreases with its atomic mass, and it is to be expected that the nucleus of the light elements should be smaller than for the heavy atoms. It is thus clear that the volume occupied by the nucleus is exceedingly small compared with that occupied by the atom as a whole.

A direct collision of an α-particle with this minute nucleus is thus a rare occurrence. It can be estimated that even in the case of heavy elements only one α-particle in about 10,000 makes a close collision with the nucleus. On account of the powerful repulsive field of the latter, the α-particle may either be turned back before reaching the nucleus, or be so diminished in energy that it is unable to effect its disruption. The case of the lighter elements, however, is much more favourable; for the repulsive forces are so much weaker that we may expect the α-particle to enter the nucleus structure without much loss of energy, and thus to be an effective agent in promoting the disintegration of the atom.

One of the most interesting cases to consider is that in which an α-particle (helium nucleus) collides with the nucleus of a hydrogen atom. Marsden showed that in some cases the *H*-atom is set in such swift motion that it can be detected by the scintillation produced on a zinc sulphide screen. The maximum speed obtainable is 1·6 times that of the incident particle, and such a swift *H*-atom is able to travel four times as far as the α-particle before being stopped. For example, the maximum range of a *H*-atom set in motion by an α-particle from radium C—range 7 cm. in air—corresponds to 29 cm. of air.

A close examination of the production of swift *H*-atoms by this method showed that the number was about 30 times greater than that to be expected if the colliding nuclei behaved as point charges repelling each other according to the inverse square law. This, and other observations, show that the law of the inverse square ceases to hold in such intense collisions, where the closest distance of approach is of the order 3×10^{-13} cm. It is probable that this distance is comparable with the actual dimensions of the structure of the α-particle itself. Some recent experiments by Chadwick and Bieler indicate that there is an abrupt change in the law of force at distances of about 5×10^{-13} cm. So far, no definite evidence has been obtained as to the nature of these forces which arise in the close collisions between nuclei. Attention should be directed to the enormous intensity of the electrical forces that come into play in such close collisions—forces much greater than can be brought to bear on an atom by ordinary laboratory methods. Unless the nucleus is a very stable structure, it is to be anticipated that it should be greatly disturbed, if not disintegrated, under the influence of such intense forces.

We must now consider the experiments which indicate that some of the lighter elements can be disintegrated by the action of α-particles. When a stream of α-particles is passed through dry air or nitrogen, a number of scintillations are observed far beyond the range of the α-particle. These scintillations are due to swift charged particles which are bent in a magnetic field like H-atoms set in swift motion by α-particles, and which, indeed, are undoubtedly H-atoms. Since this effect is not observed in dry oxygen or carbon dioxide, it appears likely that some of the nuclei of nitrogen have been disintegrated by the action of the α-particles. Recently these experiments have been repeated by Mr Chadwick and myself under much better optical conditions for counting these comparatively weak scintillations. It has been found that, using radium C as a source of α-rays, the maximum range of the H-atoms from nitrogen atoms corresponds to 40 cm. of air, while the maximum range of the H-atoms from hydrogen, or any combination of hydrogen, is only 29 cm. under similar conditions. This result negatives the possibility that the presence of these H-atoms can be ascribed to any hydrogen contamination in the ordinary chemical sense

This observation opened up a simple method of examining other elements besides nitrogen. Experiments were made beyond the maximum range (29 cm.) of ordinary H-atoms, so as to be quite independent of the presence of free or combined hydrogen in the material under examination. In this way it has been found that similar particles are produced in boron, fluorine, sodium, aluminium, and phosphorus. No definite effect has so far been observed for other elements of the production of particles with ranges greater than 29 cm. of air. The question of the production of slower velocity H-atoms has not so far been examined. The range of penetration of the atoms from aluminium is specially marked, being more than 80 cm. While no definite proof has yet been obtained of the nature of these ejected particles, it seems probable that they are in all cases H-atoms liberated from the nuclei of the elements in question. It is of special interest to note that H-atoms are only liberated in elements whose mass is given by $4n + 2$ or $4n + 3$ where n is a whole number. No H-atoms are observed from elements like carbon, oxygen, and sulphur, whose mass is given by $4n$. This is an indication that the α-particles are unable to liberate H-atoms from elements composed entirely of helium nuclei, but are able to do so from some elements composed of H-atoms as units as well as helium nuclei. It would appear as if the H-atoms were satellites of the main nuclear system and that one of them gained sufficient energy from a collision with an α-particle to escape from its orbit with a high speed. If the long-range particles from aluminium are *H*-atoms, it can be calculated that the maximum energy of motion is somewhat greater than that of the incident α-particle, indicating that the escaping fragment of the atom has gained energy from the system. It is of special interest to note that, in the case of aluminium, the direction of escape of the *H*-atom is to some extent independent of the direction of the α-particle. Nearly as many are shot in the backward as in the forward direction, but in the former case the average velocity is somewhat smaller. No element of mass greater than phosphorus (31) has been found to yield *H*-atoms. It would appear as if the constitution of the nucleus undergoes some marked change at this stage.

It should be remarked that the disintegration observed in these experiments is on a very minute scale. Only about one α-particle in a million is able to get close enough to a nucleus to effect its disintegration.

So far we have only been able to observe those fragments of atoms which escape with sufficient speed to travel further than the α-particles. Another very important method of examining the effects produced within the range of the α-particle has been recently examined by Mr Shimizu. This depends on the discovery of Mr C. T. R. Wilson that the tracks of ionising radiations can be made visible by sudden expansion of a moist gas, so that each ion becomes the centre of a visible globule of water. Wilson had previously observed an occasional bend in the track of an α-particle, with a short spur attached, indicating the collision of an α-particle with an oxygen or nitrogen nucleus. By taking a large number of photographs of tracks of α-particles, Mr Shimizu found a number of cases in which the track of the α-particle near the end of its range showed two nearly equal forks. It can readily be shewn from the range and angle between the forks that these effects cannot be ascribed to a collision of the α-particle with a *H*-atom, or with a nucleus of hydrogen or nitrogen. It would appear not unlikely that these forks indicate an actual disruption of the atom in which a helium nucleus is released. While this conclusion is only tentative, it will be of great interest to follow up further this new method of attack of a fundamental problem. It is remarkable that while only one α-particle in a million is able to liberate a *H*-atom from nitrogen, about one α-particle in 300 appears to show a forked track, indicating that this type of disintegration occurs much more frequently than the liberation of a *H*-atom.

If this interpretation proves to be correct, it shows that the amount of energy required to liberate a helium nucleus from a complex nucleus of a light atom is not great. Such a result is not inconsistent with modern ideas of the relation between mass and energy, for the fact that the atomic masses of carbon and oxygen are very nearly integral multiples of the mass of the helium atom is an indication that the helium nuclei are bound loosely together. On the other hand, if we suppose the helium nucleus itself to be composed of four hydrogen nuclei and two electrons, the loss of mass in the structure indicates that the helium nucleus is so stable a structure that it should not be dissociated by even the swiftest α-particle. This conclusion is supported by experimental observations as far as they have gone.

III. THE THEORY OF EVOLUTION: ARISTOTLE

PERHAPS an earlier subject of thought with both races and individuals than either the structure of the universe or that of matter is the origin and development of living creatures. There are stories of animal creation in every mythology; Sir J. G. Frazer writes that among primitive peoples they are of two kinds. Tillers of the soil, accustomed to the yearly flowering of fertile Earth, hold that the Creator modelled man and the animals from Earth's clay, and filled them with the breath of life. But pastoral peoples and hunters, who feel the close relationship of human and animal bodies, believe that one form of life changed into another, animals into other animals, or animals into men, so that tribes claim descent, protection, and patronymic from raven, salmon, or beaver ancestors. In historic times, these two patterns of thought alternately swayed the minds of men, till the scale was finally weighted by Darwin against the special creation theory, on the side of evolution, or the origin of higher species of living beings by descent from lower species.

Apart from the early myths, the philosophies of the Greeks are the first attempts to grapple with the problem. They are speculations only, but they tend in several cases to regard Nature as the result of a gradual development still in progress. Aristotle had some conception of a genetic series from lower to higher forms of life, and discusses the possible origin of animals either from an egg or a grub, concluding that the latter is more probable, since eggs always derive from some animal, while grubs are sometimes, as he erroneously thought, spontaneously produced. His theories have claim to remembrance, since he was a competent and thorough biologist, who observed, dissected, and described many species of animals.

HISTORIA ANIMALIUM

Book VIII. I. 588 *a*. I. (Oxford translation, 1910.)

We have now discussed the physical characteristics of animals and their modes of generation. Their habits and their modes of living vary according to their character and their food.

In the great majority of animals there are traces of psychical qualities or attitudes, which qualities are more markedly differentiated in the case of human beings. For just as we pointed out resemblances in the physical organs, so in a number of animals we observe gentleness or fierceness, mildness or cross temper, courage or timidity, fear or confidence, high spirit or low cunning, and, with regard to intelligence, something equivalent to sagacity. Some of these qualities in man, as compared with the corresponding qualities in animals, differ only quantitatively: that is to say, a man has more or less of this quality, and an animal has more or less of some other; other qualities in man are represented by analogous and not identical qualities: for instance, just as in man we find knowledge, wisdom and sagacity, so in certain animals there exists some other natural potentiality akin to these. The truth of this statement will be the more clearly apprehended if we have regard to the phenomena of childhood: for in children may be observed the traces and seeds of what will one day be settled psychological habits, though psychologically a child hardly differs for the time being from an animal; so that one is quite justified in saying that, as regards men and animals, certain psychical qualities are identical with one another, while others resemble, and others are analogous to, each other.

Nature proceeds little by little from things lifeless to animal life in such a way that it is impossible to determine the exact line of demarcation, nor on which side thereof an intermediate form should lie. Thus, next after lifeless things in the upward scale comes the plant, and of plants one will differ from another as to its amount of apparent vitality; and, in a word, the whole genus of plants, whilst it is devoid of life as compared with an animal, is endowed with life as compared with other corporeal entities. Indeed, as we have just remarked, there is observed in plants a continuous scale of ascent towards the animal. So, in the sea, there are certain objects concerning which one would be at a loss to determine whether they are animal or vegetable. For instance, certain of these objects are fairly rooted, and in several cases perish if detached; thus the pinna is rooted to a particular spot, and the solen (or razorshell) cannot survive withdrawal from its burrow. Indeed, broadly speaking, the entire genus of testaceans have a resemblance to vegetables, if they be contrasted with such animals as are capable of progression.

In regard to sensibility, some animals give no indication whatsoever of it, whilst others indicate it but indistinctly. Further, the substance of some of these intermediate creatures is fleshlike, as is the case with the so-called tethya (or ascidians) and the acalephae (or sea-anemones); but the sponge is in every respect like a vegetable. And so throughout the entire animal scale there is a graduated differentiation in amount of vitality and in capacity for motion.

A similar statement holds good with regard to habits of life. Thus of plants that spring from seed the one function seems to be the reproduction of their own particular species, and the sphere of action with certain animals is similarly limited. The faculty of reproduction, then, is common to all alike....Some animals, like plants, simply procreate their own species at definite seasons; other animals busy themselves also in procuring food for their young, and after they are reared quit them and have no further dealings with them; other animals are more intelligent and endowed with memory, and they live with their offspring for a longer period, and on a more social footing.

The life of animals, then, may be divided into two acts—procreation and feeding; for on these two acts all their interest and life concentrate. Their food depends chiefly on the substance of which they are severally constituted; for the source of their growth in all cases will be this substance. And whatsoever is in conformity with nature is pleasant, and all animals pursue pleasure in keeping with their nature.

DE GENERATIONE ANIMALIUM

BOOK III. II. 762 *b*. 28. (Oxford translation, 1911.)

(Hence) one might suppose, in connection with the origin of men and quadrupeds, that, if ever they were really "earth-born" as some[1] say, they came into being in one of two ways; that either it was by the formation of a scolex[2] at first or else it was out of eggs. For either they must have had in themselves the nutriment for growth (and such a conception is a scolex) or they must have got it from elsewhere, and that either from the mother[3] or from part of the conception[4]. If then the former is impossible (I mean that nourishment should flow to them from the earth as it does in animals from the mother), then they must have got it from some part of the conception, and such generation we say is from an egg[5].

It is plain then that, if there really was any such beginning of the generation of all animals, it is reasonable to suppose it to have been one of these two, scolex or egg. But it is less reasonable to suppose that it was from eggs, for we do not see such generation occurring with any animal[6], but we do see the other both in the sanguinea above mentioned[7] and in the bloodless animals.

[1] Anaximander, who said the first animals sprang from the slime of the sea.

[2] scolex = grub or embryo, defined earlier by Aristotle as an egg the whole of which develops into the animal, i.e. without a supply of nutriment like the yolk of an egg.

[3] as in vivipara.

[4] as in ovipara.

[5] therefore if they got it from elsewhere at all they must have begun in the form of an egg, not of a complete organism, but we shall see directly that this view is also unlikely.

[6] we do not see eggs spontaneously produced.

[7] mullets and eels; according to Aristotle, eels are actually developed from a scolex—the earth-worm.

THE DARK AGES: PLINY

ARISTOTLE's philosophic speculations were gradually lost sight of, and for many hundred years were forgotten in Europe. Some of his observations and collected facts did, however, enter into the composition of a book that was read throughout the Middle Ages—Pliny's *Natural History*. This work, in which the heavens, the earth, man, animals, plants, agriculture, fine arts, and black magic are passed in review, is an attempt to combine into a connected whole the knowledge and beliefs of a series of Greek and Roman writers. The author, called the elder Pliny, was born in North Italy in 29 A.D., and after fighting in German campaigns, became a pleader in the law courts of Rome, and a diligent student of Greek and Roman literature. At the time of his death, he was in command of the Roman fleet, and perished through venturing on land to watch the great eruption of Vesuvius that overwhelmed Pompei and Herculaneum in 79 A.D.

The spirit of this fatal exhibition of scientific curiosity is little in evidence in his book. Francis Bacon writes in 1605: "...in natural history, we see there hath not been that choice and judgment used as ought to have been; as may appear in the writings of Plinius, Cardanus, Albertus and divers of the Arabians; being fraught with much fabulous matter, a great part not only untried, but notoriously untrue, to the great derogation of the credit of natural philosophy with the grave and sober kind of wits. Wherein the wisdom and integrity of Aristotle is worthy to be observed; that having made so diligent and exquisite a history of living creatures, hath mingled it sparingly with any vain or feigned matter; and yet on the other side hath cast all prodigious narrations which he thought worthy the recording into one book; excellently discerning that matter of manifest truth, such whereupon observation and rule was to be built, was not to be mingled or weakened with matter of doubtful credit; and yet again that rarities and reports that seem uncredible are not to be suppressed or denied to the memory of men."

We quote from an early seventeenth century translation of Pliny.

THE HISTORY OF THE WORLD

Commonly called The Natural History of

C. PLINIUS SECUNDUS.

Translated into English by Philemon Holland, Doctor of Physicke.

London, Printed by *Adam Islip*, 1634.

The Seventh Booke. *The Proeme.*

Thus, as you see, we have in the former books sufficiently treated of the universall world; of the Lands, Regions, Nations, Seas, Islands, and renowned Cities therein contained. It remaines now to discourse of the living creatures comprised within the same, and their natures: a point doubtlesse that would require as deepe a speculation as any part else thereof whatsoever, if so be the spirit and minde of man were able to comprehend and compasse all things in the world. And to make a good entrance into this treatise and history, me thinkes of right we ought to begin at Man, for whose sake it should seeme that Nature made and produced all other creatures besides, though this great favour of hers, so bountifull and beneficiall in that respect, hath cost them full deare. Insomuch as it is hard to judge, whether in so doing she hath done the part of a kind mother, or a hard and cruell step-Dame. For first and foremost, of all other living creatures, man she hath brought forth all naked, and cloathed him with the good and riches of others. To all the rest she hath given sufficient to clad them every one according to their kinde; as namely, shells, cods, hard hides, prickes, shag, bristles, haire, downe feathers, quills, skales, and fleeces of wooll. The very trunkes and stems of trees and plants she hath defended with barke and rinde, yea and the same sometimes double, against the injuries of heate and cold: Man alone, poore wretch, she hath layed all naked upon the bare earth, even on his birthday; to cry and wraule presently from the very first houre that hee is borne in such sort, as among so many living creatures there is none subject to shed teares and weepe like him. And verily to no babe or infant is it given to laugh before it be forty daies old, and that is counted very early, and with the soonest. Moreover, so soone as he is entred in this manner to enjoy the light of the Sunne, see how he is immediately tyed and bound fast, and hath no member at libertie: a thing that is not practised upon the yong whelpes of any beast among us, be he never so wilde.... How long is it ere we can go alone? how long before we can prattle and speake, feed ourselves, and chew our meat strongly?... What should I say of the infirmities and sicknesses that do soone seise upon our feeble bodies? what need I speake of so many medicines and remedies devised against these maladies: besides the new diseases that come every day, able to checke and frustrate all our provision of physicke whatsoever? As for all other living creatures, there is not one, but by a secret instinct of nature knoweth his own good, and wherto he is made able; some make use of their swift feet, others of their flight wings; some are strong of limne; others are apt to swim, and practise the same: man only knoweth nothing unlesse he be taught; hee can neither speake, nor goe, nor eate, otherwise than he is trained to it: and to be short, apt and good at nothing he is naturally, but to pule and cry....To conclud, all other living creatures live orderly and well, after their owne kinde: we see them flocke and gather together, and ready to make head and stand against all others of a contrary kinde: the lyons as fell and savage as they be, fight not with one another: serpents sting not serpents, nor bite one another with their venomous teeth: nay the very monsters and huge fishes of the sea, war not among themselves in their owne kinde: but beleeve me, Man at mans hand receiveth most harme and mischiefe.

There follow LX chapters on the characteristics of men of different countries, on individuals remarkable for various reasons, on accidents of birth, life, and death, and other matters.

THE EIGHTH BOOKE. Chap. I. *Of landbeasts. The praise of Elephants: their wit and understanding.*

Passe we now to treat of other living creatures, and first of landbeasts: among which, the Elephant is the greatest, and commeth neerest in wit and capacitie, to men; for they understand the language of that country wherin they are bred, they do whatsoever they are commanded, they remember what duties they be taught, and withall take a pleasure and delight both in love and also in glory, nay more than all this, they embrace goodnesse, honestie, prudence, and equitie (rare qualities I may tel you to be found in men) and withall have in religious reverence (with a kinde of devotion) not only the stars and planets, but the sun and moon they also worship. And in very truth, writers there be who report thus much of them, That when the new moon beginneth to appear fresh and bright, they come downe by whole heards to a certaine river named Amelus, in the desarts and forests of Mauritania, where after that they are washed and solemnly purified by sprinckling and dashing themselves all over with water, & have saluted and adored after their manner that planet, they returne again into the woods and chases, carrying before them their yong calves that be wearied and tired. Moreover, they are thought to have a sense and understanding of religion & conscience in others; for when they are to passe the seas into another country, they will not embarke before they be induced thereto by an oath of their governors and rulers, That they shall returne again: and seene there have bin divers of them, being enfeebled by sicknesse (for as big and huge as they be, subject they are to grievous maladies) to lie upon their backs, casting and flinging herbes up toward heaven, as if they had procured and set the earth to pray for them. Now for their docility and aptnesse to learne any thing; the king they adore, they kneele before him, and offer unto him garlands and chaplets of floures and green herbes. To conclude, the lesser sort of them, which they call Bastards, serve the Indians in good stead to eare and plough their ground.

There follow LIX chapters on four-footed animals of all kinds.

THE NINTH BOOKE. Chap. I. *The Nature of water Creatures.*

I have thus shewed the nature of those beasts that live upon the land, and therin have some societie and fellowship with men. And considering, that of all others besides in the world, they that flie be the least, we will first treat of those fish that keep in the sea, not forgetting those also either in running fresh rivers or standing lakes.

Chap. II. *What the reason is why the sea should breed the greatest living creatures.*

The waters bring forth more store of living creatures, and the same greater than the land. The cause wherof is evident, even the excessive abundance of moisture. As for the fouls & birds, who live hanging, as it were, & hovering in the aire, their case is otherwise. Now the sea, being so wide, so large and open, readie to recive from heaven above the causes of generation; being so soft and pliable, so proper & fit to yield nourishment and encrease; assisted also by Nature, which is never idle, but alwaies framing one new creature or other: no marvell is it if there are found so many strange and monstrous things as there be. For the seeds and universall elements of the world are so interlaced sundry waies, and mingled one within another, partly by the blowing of the winds, and partly with the rolling and agation of the waves, insomuch as it may truly be said, according to the vulgar opinion, that whatsoever is engendred and bred in any part of the world besides, is to be found in the sea: and many more things in it, which no where else are to be seen. For there shall ye meet with fishes, resembling not onely the forme and shape of land creatures living, but also the figure and fashion of many things without life: there may one see bunches of grapes, swords, and sawes, represented; yea, and also cowcumbers, which for colour, smell, and taste, resembleth those growing upon the earth. And therefore we need the less to wonder, if in so little shell fishes as are cockles, there be somewhat standing out like horse-heads.

There follow:

Chap. III. *Of the monstrous fishes in the Indian sea.*
Chap. IIII. *Which be the greatest fishes in any coast of the Ocean sea.*
Chap. V. *Of Tritons, Nereides, and sea-Elephants, and their formes.*
Chap. VI. *Of the Balænes and Orcæ.*

Chap. VII. *Whether fish do breath and sleep, or no.*

All writers are fully resolved in this, That the Whales abovesaid, as well the Balænæ as the Orcæ, and some few other fishes bred & nourished in the sea, which among other inward bowels have lungs, doe breath. For otherwise it were not possible, that either they or any other beast, without lungs should blow: and they that be of this opinion, suppose likewise, that no fishes having guils, do draw in and deliver their wind again to and fro: nor many other kinds besides, although they want the foresaid gils. Among others, I see that *Aristotle* was of that mind, and by many profound and learned reasons persuaded & induced many more to hold the same. For mine own part, if I should speak frankly what I think, I professe that I am not of their judgment. For why? Nature if she be so disposed, may give instead of lungs some other organs and instruments of breath: to this creature one, to that another: like as many other creaturs have another kind of moist humor in lieu of blood. And who would marvel, that this vitall spirit should pierce within the waters, considering that he seeth evidently how it riseth againe and is delivered from thence: also how the aire entreth even into the earth, which is the grosest & hardest of al the elements? As we may perceive by this good argument, that some creatures, which albeit they be alwaies covered within the ground, yet live and breath neverthelesse, and namely, the Wants or Mold-warpes. Moreover, I have divers pregnant & effectuall reasons inducing me to beleeve, that all water creatures breathe each one after their maner, as Nature hath ordained. First and principally, I have observed oftentimes by experience, That fishes evidently breath and pant for wind (after a sort) in the great heat of Summer: as also that they yawne and gape when the weather is calme & the sea still. And they themselves also who hold the contrarie, confesse plainly, that fishes doe sleepe. And if that be true, How, I pray you, can they sleep if they take not their wind? Moreover, whence come those bubbles which continually are breathed forth from under the water? and what shall we say to those shell fishes which wax and decay in substance of bodie, according to the effect of the Moones encrease or decrease? But above all, fishes have hearing and smelling, and no doubt both these senses are performed and maintained by the benefit and matter of the aire: for what is smell and sent, but the verie aire, either infected with a bad, or perfumed with a good favour? Howbeit I leave every man free to his own opinion, as touching these points. But to return again to our purpose: this is certaine, that neither the Whales called Balænæ, nor the Dolphins, have any guills: and yet do both these fishes breathe at certaine pipes and conduits, as it were reaching down into their lungs: from the forehead, in the Balænes; and in the Dolphins, from the backe. Furthermore, the Sea-calves or seales, which the Latines call Phocæ, doe both breath and sleepe upon the drie land. So do the sea Tortoises also, whereof we will write more anon.

There follow LV chapters on water-creatures, including Dolphins, Tortoises, Seals, Oysters, Eels, and Fishes of all kinds.

THE TENTH BOOKE. Chap. I. *The nature of Birds and Foules.*

It followeth now that we should discourse of the nature of Foules. And first to begin with Ostriches. They are the greatest of all other foules, and in manner of the nature of foure footed beasts: (namely, those in Africke and Æthiopia) for higher they be than a man sitting on horsebacke is from the ground: and as they be taller than the man, so are they swifter on foot than the very horse: for to this end only hath Nature given them wings, even to help and set them forward in their running: for otherwise, neither flie they in the aire, ne yet so much as rise & mount from the ground. Cloven hoofs they have like red deere, and with them they fight; for good they be to catch up stones withall, & with their legs they whurle them back as they run away, against those that chase them. A wonder is this in their nature, that whatsoever they eat (and great devourers they be of al things, without difference and choice) they concoct and digest it. But the veriest fooles they be of all others. For as high as the rest of their body is, yet if they thrust their head and necke once into any shrub or bush, and get it hidden, they thinke then they are safe enough, and that no man seeth them. Now two things they doe afford, in recompence of mens pains that they take in hunting and chasing them: to wit, their egs, which are so big, that some use them for vessels in the house: and their feathers so faire, that they serve for pennaches to adorne and set out the crests and morions of souldiers in the wars.

Chap. II. *Of the Phœnix.*

The birds of Æthiopia and India, are for the most part of divers colours, and such as a man is hardly able to decipher and describe, But the Phœnix of Arabia passes all others. Howbeit, I cannot tell what to make of him: and first of all, whether it be a tale or no, that there is never but one of them in all the world, & the same not commonly seen. By report he is as big as an Ægle: for colour, as yellow and bright as gold (namely all about the necke;) the rest of the bodie a deep red purple: the taile azure blew, intermingled with feathers among of rose carnation colour; and the head bravely adorned with a crest and penach finely wrought; having a tuft and plume thereupon, right faire and goodly to be seen. *Manilius*, the noble Romane Senatour, right excellently seene in the best kind of learning and litterature, and yet never taught by any, was the first man of the long Robe, who wrot of this bird at large, & most exquisitely He reporteth, that never man was known to see him feeding: that in Arabia he is held a sacred bird, dedicated unto the Sun: that he liveth 660 yeares: and when he groweth old, and begins to decay, he builds himselfe a nest with the twigs and branches of the Canell or Cinamon, and Frankincense trees: and when he hath filled it with all sort of sweet Aromaticall spices, yeeldeth up his life thereupon. He saith moreover that of his bones and marrow there breedes at first as it were a little worme: which afterwards prooveth to be a prettie bird. And the first thing that this yong new Phœnix doth, is to perform the obsequies of the former Phœnix late deceased: to translate and cary away his whole nest into the citie of the Sun neere Panchea, and to bestow it full devoutly there upon the altar. The same *Manilius* affirmeth, that the revolution of the great yeare so much spoken of, agreeth just with the life of this bird: in which yeare the stars returne again to their first points, and give significations of times and seasons, as at the beginning and withall, that this yeare should begin at high noone that very day when the Sun entreth the signe Aries. And by his saying, the yeare of that revolution was by him shewed, when *P. Licinus* and *M. Cornelius* were consuls, *Cornelius Valerianus* writeth, That whiles *Q. Plantius* and *Sex. Papinius* were Consuls, the Phœnix flew into Ægypt.

Brought was he hither also to Rome in the time that *Claudius Cæsar* was Censor, to wit, in the eight hundreth yeare from the foundation of Rome: and shewed openly to be seen in a full hall and generall assembly of the people, as appeareth upon the publick records: howbeit, no man ever made any doubt, but he was a counterfeit Phœnix, and no better.

There follow LXXIII chapters on birds.

The next book treats of insects, small animals, and various parts of the animal body.

MEDIAEVAL ALLEGORIES: PHYSIOLOGUS

EVEN Pliny's fables pale beside the extravagance of the mediaeval allegories. The mediaeval schoolmen wished all sciences to be handmaids of theology; we may feel that in the next extract, which is typical of mediaeval writings, zoology occupies rather the position of a slave.

Physiologus was the common title of a collection of fifty Christian allegories, probably compiled in Alexandria in the very early centuries of the Christian era. It was read all over Europe and the Near East. The lion, the elephant, the phoenix and many others figure therein as the texts for commentaries on Scripture.

PHYSIOLOGUS OF LEYDEN

(From the translation by P. N. Laud. Leiden, 1875.)

Chapter One. *Of the lion, who is king of beasts.*

We begin with the lion, who is called the king of beasts. For he is an image of the Lord our God, as in the saying of Blessed Jacob, who, when he praised Juda, prophesied and said: "Juda is a lion's cub; from the prey, my son, art thou gone up." Physiologus speaks of the same lion in this way: three qualities are inherent in his nature, and his nature is governed by three laws.

This is the first quality. When he wanders to and fro and travels in the mountains, he smells the hunters afar off; and lest they should find his footprints and come after him, he flattens the footprints with his tail and covers them up, so that the hunters do not see his tracks and rise up against his hiding-place and take him. *Exposition.* Even so, the Lord God our Saviour, the most noble and undefiled lion of the tribe of Juda and the race of Israel and the branch of David, when he was sent by the Eternal

Father, was born corporeally, and concealed his most noble footprints, i.e., his divinity, by the sacrosanct body of human origin existing for human salvation.

The second quality of the same lion, king of beasts. His second quality, or custom, is of this nature. The lion, when he sleeps, has open and watchful eyes. Of the same mystery the holy Church in the Song of Songs spoke to her spouse, when rejoicing she exclaimed: I sleep and my heart watches; i.e., the body of Christ slept, who is the head of the Church, and who by his divine will slept the quiet sleep of bodily death; the deity of the same Word, which is the mind and heart of the holy Church, watched and did not sleep in death, being God and immortal. For the Church is the body of Christ, as the apostolic writings teach, and God the Word with the Father and the Holy Spirit is her mind and understanding and faith.

The third quality of the same lion, or his power of generation. The third quality of the lion is of this kind. When the lioness bears her cubs, they are born dead; but she watches over them until the third day, when the father is accustomed to come to them. When he comes and finds them dead, he breathes between their eyes and makes them arise. *Theory.* In the same way God the Omnipotent Father made arise on the third day the firstborn of all creatures, namely our Lord Jesus, his beloved son and the true God. And also our Lord after he had risen from death, breathed into his holy disciples; and raised up Adam again into the communion of the Holy Spirit, which he had lost when he transgressed the command. Very clearly spoke Jacob: He couched as a lion's cub and who shall rouse him up. *Theory of another kind.* This is otherwise expounded by some, who say of the lion and the lion's cub that they are like those who overcome themselves, yet do not endure to the end of their labours, those who of their own wish cloister themselves and are dead to the world and its ugly sins. Then there comes to them the father of sins and inventor of perdition and defection from God, who is like a lion, gigantic in evil; he finds them dead to the world and remote from his communion of sinfulness, and he is concerned for them, and comes to the place, and breathes his bitterness into them, and makes them live to sin and the world, to which they were dead and from which they had escaped by good works and the propinquity of God.

HOOKE: AN EARLY MICROSCOPIST

LIKE all other sciences, biology awoke at the time of the Renaissance, re-read Aristotle, and settled to a new era of patient observation. Striking advances were few; we find in the notebooks of Leonardo da Vinci the foreshadowing of later discoveries in anatomy, physiology, and optics; Andreas Vesalius published a book on human anatomy, based on actual dissection, not, as were previous works, on what Galen taught; William Harvey made clear the mechanism of the circulation of the blood. In the seventeenth century the improvement of the microscope gave an impetus to biological observation like that imparted to astronomy by Galileo's telescope. No one man was responsible for this improvement, but perhaps the two to whom it was chiefly due were Anthony van Leeuwenhoek (1632–1723), a glass-grinder of Delft, who was elected a member of the Royal Society of London as a result of his work on microscopes, and Robert Hooke, who lived from 1635 to 1703. The latter was brought up in Sir Peter Lely's workshop, and was employed, after graduating at Oxford, by Robert Boyle, the chemist. He was curator of experiments to the Royal Society for 41 years, and secretary for five years. Both he and Leeuwenhoek published collections of observations; we quote from Hooke's *Micrographia* a typical account of the Gnat, with speculations on spontaneous generation. Besides his microscopic work, Hooke devised methods of flying, improved the construction of watches, made plans and acted as surveyor for the rebuilding of London after the Great Fire, suggested weather forecasts, and approached the theory of gravitation. His very varied knowledge thus led him to originate much, but perfect little. Owing to his solitary and irritable nature, he had a lonely life and few friends.

MICROGRAPHIA

Or some PHYSIOLOGICAL DESCRIPTIONS of MINUTE BODIES made by MAGNIFYING GLASSES. With Observations and enquiries thereupon.

By R. HOOKE, Fellow of the Royal Society.
London, 1665.

Observ. XLIII. *Of the* WATER-INSECT *or* GNAT.

THIS little creature, described in the first *Figure* of the 27. *Scheme*, was a small scaled or crusted Animal which I have often observ'd to be generated in Rain-water; I have also observ'd it both in Pond and River-water. It is suppos'd by some, to deduce its first original from the putrifaction of Rain-water, in which, if it have stood any time open to the air, you shall seldom miss, all the Summer long, of store of them striking too and fro.

'Tis a creature, wholly differing in shape from any I ever observ'd; nor is its motion less strange: It has a very large head, in proportion to its body, all covered with a shell, like other *testaceous* Animals, but it differs in this, that it has, up and down several parts of it, several tufts of hairs, or bristles, plac'd in the order express'd in the Figure; It has two horns, which seem'd almost like the horns of an Oxe, inverted, and, as neer as I could guess, were hollow, with tufts of brisles, likewise at the top: these horns they could move easily this or that way, and might perchance, be their nostrils. It has a pretty large mouth, which seem'd contriv'd much like those of Crabs and Lobsters, by which, I have often observ'd them to feed on water, or some imperceptible nutritive substance in it.

I could perceive, through the transparent shell, while the Animal surviv'd, several motions in the head, thorax, and belly, very distinctly, of differing kinds which I may, perhaps, elsewhere endeavour more accurately to examine, and to shew of how great benefit the use of a *Microscope* may be for the discovery of Nature's course in the operations perform'd in Animal bodies, by which we have the opportunity of observing her through these delicate and pellucid teguments of the bodies of Insects acting according to her usual course and way, undisturbed, whereas, when we endeavour to pry into her secrets by breaking open the doors upon her, and dissecting and mangling creatures while there is life yet within them, we find her indeed at work, but put into such disorder by the violence offer'd, as it may easily be imagin'd, how differing a thing we should find, if we could, as we can with a *Microscope*, in these smaller creatures, quietly peep in at the windows, without frighting her out of her usual byas.

The form of the whole creature, as it appear'd in the *Microscope*, may, without troubling you with more descriptions, be plainly enough perceiv'd by the *scheme*, the hinder part or belly consisting of eight several jointed parts, namely, *ABCDEFGH*, of the first *Figure*, from the midst of each of which, on either side, issued out three or four small bristles or hairs, *I*, *I*, *I*, *I*, *I*, the tail was divided into two parts of very differing make; one of them, namely, *K*, having many tufts of hair or bristles, which seem'd to serve both for the finns and tail, for the Oars and Ruder of this little creature, wherewith it was able, by striking

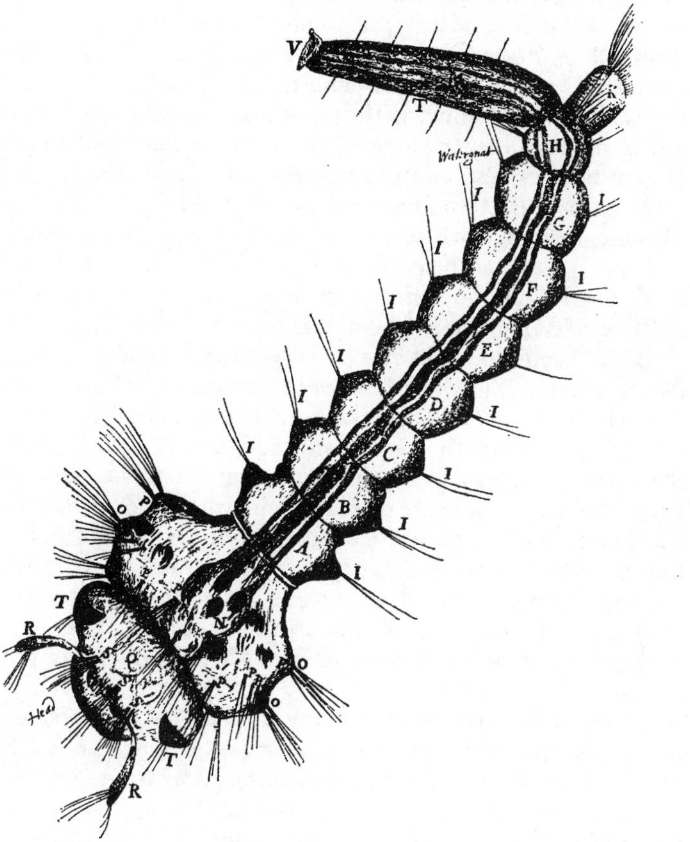

Fig. 1.

and bending its body nimbly to and fro, to move himself any whither, and to skull and steer himself as he pleas'd; the other part, *L*, seem'd to be, as 'twere, the ninth division of his belly, and had many single bristles on either side. From the end *V*, of which, through the whole belly, there was a kind of Gut of a darker colour, *MMM*, wherein, by certain *Peristaltick* motions there was a kind of black substance mov'd upwards and downwards through it from the orbicular part of it, *N*, (which seem'd the *Ventricle*, or stomach) to the tail *V*, and so back again, which *peristaltick* motion I have observ'd also in a Louse, a Gnat, and several other kinds of transparent body'd Flies. The *Thorax* or chest of this creature *OOOO*, was thick and short, and pretty transparent, for through it I could see the white heart (which is the colour also of the blood in these, and most other Insects) to beat, and several other kind of motions. It was bestuck and adorn'd up and down with several tufts of bristles, such as are pointed out by *P*, *P*, *P*, *P*, the head *Q* was likewise bestuck with several of those tufts, *SSS*; it was broad and short, had two black eyes, *TT*, which I could not perceive at all pearl'd, as they afterwards appear'd, and two small horns, *RR*, such as I formerly describ'd.

Both its motion and rest is very strange, and pleasant, and differing from those of most other creatures I have observ'd; for, where it ceases from moving its body, the tail of it seeming much lighter than the rest of its body, and a little lighter than the water it swims in, presently boys it up to the top of the water, where it hangs suspended with the head always downward; and like our *Antipodes*, if they do by a frisk get below that superficies, they presently ascend again into it, if they cease moving, until they tread, as it were, under that superficies with their tails; the hanging of these in this posture, put me in mind of a certain creature I have seen in *London*, that was brought out of *America*, which would very firmly suspend it self by the tail, with the head downwards, and was said to sleep in that posture, with her young ones in her false belly, which is a Purse, provided by Nature for the production, nutrition, and preservation of the young ones, which is describ'd by *Piso* in the 24. Chapter of the Fifth Book of his Natural History of *Brazil*.

The motion of it was with the tail forwards, drawing its self backwards, by the striking to and fro of that tuft which grew out of one of the stumps of its tail. It had another motion, which was more sutable to that of other creatures, and that is, with the head forward; for by the moving of his chaps (if I may so call the parts of his mouth) it was able to move it self downwards very gently towards the bottom, and did, as 'twere, eat up its way through the water.

But that which was most observable in this creature, was, its Metamorphosis or change; for having kept several of these Animals in a Glass of Rain-water, in which they were produc'd, I found, after about a fortnight or three weeks keeping, that several of them flew away in Gnats, leaving their hulks behind them in the water floating under the surface, the place where these Animals were wont to reside, whil'st they were inhabitants of the water; this made me more diligently to watch them, to see if I could find them at the time of their transformation; and not long after, I observ'd several of them to be changed into an unusual shape, wholly differing from that they were of before, their head and body being grown much bigger and deeper, but not broader, and their belly, or hinder part smaller, and coyl'd, about this great body much of the fashion represented by the prick'd line in the second *Figure* of the 27 *Scheme*, the head and horns now swam uppermost, and the whole bulk of the body seem'd to be grown much lighter; for when by my frighting of it, it would by striking out of its tail... sink it self below the surface towards the bottom; the body would more swiftly reascend, than when it was in its former shape.

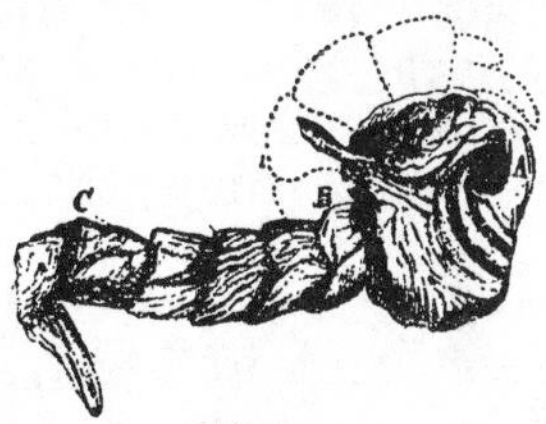

Fig. 2.

I still marked its progress from time to time, and found its body still to grow bigger and bigger, Nature, as it were, fitting and accoutring it for the lighter Element, of which it was now going to be an inhabitant; for, by observing one of these with my *Microscope*, I found the eyes of it to be altogether differing from what they seem'd before, appearing now all over pearl'd or knobb'd, like the eyes of Gnats, as is visible in the second *Figure* by A. At length, I saw part of this creature to swim above, and part beneath the surface of the water, below which though it would quickly plunge it self if I by any means frighted it, and presently re-ascend into its former posture; after a little longer expectation, I found that the head and body of a Gnat, began to appear and stand cleer above the surface, and by degrees it drew out its leggs, first the two formost, then the other, at length its whole body perfect and entire appear'd out of the husk (which it left in the water) standing on its leggs upon the top of the water, and by degrees it began to move, and after flew about the Glass a perfect Gnat.

I have been the more particular, and large in the relation of the transformation of divers of these little Animals which I observ'd, because I have not found that any Authour has observ'd the like; and because the thing it self is so strange and heterogeneous from the usual progress of other Animals, that I judge it may not onely be pleasant, but very usefull and necessary towards the completing of Natural History...

But to return to the more immediate consideration of our Gnat: We have in it an instance, not usual or common, of a very strange *amphibious* creature, that being a creature that inhabits the Air, does yet produce a creature, that for some time lives in the water as a Fish, though afterward (which is as strange) it becomes an inhabitant of the Air, like its Sire, in the form of a Fly. And this, me thinks, does prompt me to propose certain conjectures, as Queries, having not yet had sufficient opportunity and leisure to answer them my self from my own Experiments or Observations.

And the first is, Whether all those things that we suppose to be bred from corruption and putrifaction, may not be rationally suppos'd to have their origination as natural as these Gnats, who, 'tis very probable, were first dropt into this Water, in the form of Eggs. Those Seeds or Eggs must certainly be very small, which so small a creature as a Gnat yields, and therefore we need not wonder that we find not the Eggs themselves, some of the younger of them, which I have observ'd, having not exseeded a tenth part of the bulk they have afterwards come to; and next, I have observed some of those little ones which must have been generated after the Water was inclosed in the Bottle, and therefore most probably from Eggs, whereas those creatures have been suppos'd to be bred of the corruption of the Water, there being not formerly known any probable way how they should be generated.

A second is, whether these Eggs are immediately dropt into the Water by the Gnats themselves, or, mediately, are brought down by the falling rain; for it seems not very improbable, but that these small seeds of Gnats may (being, perhaps, of so light a nature, and having so great a proportion of surface to so small a bulk of body) be ejected into the Air, and so, perhaps, carried for a good while too and fro in it, till by the drops of Rain it be wash'd out of it.

A third is, whether multitudes of those other little creatures that are found to inhabit the Water for some time, do not, at certain times, take wing and fly into the Air, others dive and hide themselves in the Earth, and so contribute to the increase both of the one and the other Element.

SPECIES: THE LINNAEAN SYSTEM OF CLASSIFICATION

THE progress of observation in biology was forwarded in another way by the Linnaean system of classification, enabling newly discovered species to be correlated with those already known. Born in Sweden in 1707, Carl van Linné (latinized to Carolus Linnaeus) was intended for the Church, but his interest in botany led him to scientific pursuits. After some years as a struggling student and doctor, he was helped by rich patrons, travelled widely, and ultimately became professor of botany in the University of Upsala. He died in 1708. His system of classification, founded on the sex organs of plants, held its own till replaced by the modern attempt to group organisms according to their natural relationships; in this the sex organs of plants are regarded, it is interesting to note, as the most reliable indications of descent. Linnaeus recognized that his system was not a natural one; yet he was in no way an evolutionist, and believed in the absolute fixity of species.

The doctrine of the special creation and fixity of species had become the orthodox teaching of the Church, enforced in Protestant countries by the emphasis on the Bible and the literal interpretation thereof. Yet the early Christian fathers, notably St Augustine, harmonized Greek evolutionary philosophy with the Hebrew creation stories. St Augustine writes: "even so as in the grain itself there were invisible all things simultaneously which were in time to grow into the tree, so the world itself is to be thought of when God simultaneously created all things, as having at the same time in itself all things that were made in it and with it, when the day itself was created: not only the heaven with the sun and moon and stars, and so forth, but also those things which the water and earth produced *potentialiter atque causaliter*; before that, in due time, and after long delays, they grew up in such manner as they are now known to us in those works of God which he is working even to the present hour."

But Aristotle's biology, unlike his physics, never became part of the later doctrine of the Church, and the acceptance of the literal teaching of Genesis as to the special creation of each species gradually developed into a *sine qua non* of orthodox belief. The rising tide of scientific observation and classification in the seventeenth century supported the story of Genesis at first. The task of reducing to order the myriad species of plants and animals was only rendered feasible by the conviction that each species was separate and distinct; no change in species was seen to occur, therefore no change had occurred since the evening and the morning were the sixth day. We see this view reflected in Linné's introduction to his work on the *Families of Plants*, printed first in the Netherlands in 1737.

THE FAMILIES OF PLANTS

With their NATURAL CHARACTERS, according to the NUMBER, FIGURE, SITUATION, and PROPORTION of all the parts of FRUCTIFICATION.

Translated from the...*GENERA PLANTARUM*,...of the elder LINNEUS.

By a BOTANICAL SOCIETY at Lichfield, 1787.

ACCOUNT OF THE WORK.

1. All the real knowledge, which we possess, depends on METHOD; by which we distinguish the similar from the dissimilar. The greater number of natural distinctions this method comprehends, the clearer becomes our idea of the things. The more numerous the objects, which employ our attention, the more difficult it becomes to form such a method; and the more necessary. The great Creator has in no part of his works presented a greater variety to the human mind than in the vegetable kingdom; which covers the whole globe, which we inhabit; whence, if a distinct method is ever necessary, it is necessary here; if we hope to gain a distinct knowledge of vegetables....

2. To him therefore Vegetables are known, who (1) can join the similar to the similar, and can separate the dissimilar from the dissimilar.

3. The BOTANIST is he, who can affix similar names (2) to similar vegetables, and different names to different ones, so as to be intelligible to every one.

4. The NAMES (3) of Plants are *generic*, and (where there are any species), *specific*. These should be certain and well founded, not vague, evasive, or variously applicable. Before they can be such, it is necessary, that they should have been affix'd to certain, not to vague genera (2. 6) for if this foundation be unsteady, the names also, and in consequence the doctrine of the Botanist crumbles into ruin (3).

5. The SPECIES are as numerous as the different and constant forms of vegetables, which exist upon this globe; which forms according to instinctive laws of generation produce others, similar to themselves, but in greater numbers. Hence there are as many Species, as there are different forms or structures of Plants now existing; excepting such less-different *Varieties*, which situation or accident has occasion'd.

6. The GENERA are as numerous, as the common proximate attributes of the different species, (5), as they were created in the beginning; this is confirm'd by revelation, discovery, observation, hence

THE GENERA ARE ALL NATURAL.

For we must not join in the same genus the horse and the swine, tho' both species had been one-hoof'd, nor separate in different genera the goat, the raindeer, and the elk, tho' they differ in the form of their horns. We ought therefore by attentive and diligent observation to determine the limits of the genera, since they cannot be determin'd a priori. This is the great work, the important labour, "*for should the Genera be confused, all would be confusion.*" Cesalp.

7. That it has pleased Infinite Wisdom to distinguish the Genera of Plants by their FRUCTIFICATION was discover'd in the last age; and first indeed by CONRADUS GESNER, the ornament of his time; as appears from his posthumous epistles, and from the plates published by CAMERARIUS, altho' the first, who introduced this great discovery into use, was ANDREAS CESALPINUS; which would nevertheless have shortly expired in its cradle, unless it had been recalled into life by the care of ROBERT

MORISON, and nourish'd by JOSEPH P. TOURNEFORT with pure systematic rules. This was at length confirm'd by all the great men, then existing, in the science.

8. This foundation being given (7), this point fix'd, immediately every one capable of such researches join'd their labours to turn it into use, to build a system; all with the same inclination, and to the same purpose, but not all with equal success. Because the fundamental rule was known but to few, which if the builders did not observe, quickly fell to the ground with the first tempest the insubstantial edifice, however splendid; Boerhave (Instit. 31) well observes, that *a* TEACHER, *as he explains discoveries, may proceed from generals to particulars, but an* INVENTOR *on the contrary must pass from particulars to generals.*"

For some assuming the different parts of Fructification as the principle of their system, and descending according to the laws of division from classes through orders even to species, broke and dilacerated the natural Genera (6), and did violence to nature by their hypothetical and arbitrary principles. For example, one from the *Fruit* denies that the Peach and Almond can be join'd in the same genus, another from the *regularity of the Petals* denies that Valerian and Valerianoides; another from their *number* that Flax and Radiola; another from their *sex* that the urtica androgyna, and dioica, &c. can be combined under the same genus: for, say they, if these cannot be conjoin'd in the same class, much less in the same genus: not having observed, that themselves have contrived the classes, but that the Creator himself made the Genera. Hence arose so many false Genera! such controversy amongst authors! so many bad names! such confusion! Such indeed was the state of things, that as often as a new System-maker arose, the whole Botanic world was thrown into a panic. And for my part I know not, whether these System-builders produced more evil than good to the science; this is certain, if the unlearned be compared with the learned, they much surpass them in number, Physicians, Apothecaries, Gardiners lamented this misfortune, and with reason. I confess, their theory had been excellent, had it pleased the great Creator to have made all the fructifications of the same Genus of Plants equally similar amongst themselves; as are the individuals of the same species. As this is not so, we have no resource, since we are not the governors of Nature, nor can create plants according to our own conceptions, but to submit ourselves to the laws of nature, and learn by diligent study to read the characters inscribed on plants. If every different mark of the fructification be adjudged sufficient for distinguishing the genus, why should we a moment hesitate to proclaim nearly as many genera as species? for we are scarcely acquainted with any two flowers so similar to each other, but that some difference of their parts may be discern'd; I also once endeavour'd from the flower alone to determine all the specific differences, but frequently with less success, since there is an easier method. I wish it therefore to be acknowledged by all true Botanists, if they ever expect any certainty in the science, that *the Genera and Species must be all natural*; without which assumed principle there can be nothing excellent done in the science....

9.... I do not deny, that natural Classes may not be given as well as natural Genera. I do not deny, that a natural method ought much to be prefer'd to ours, or those of other discoverers; but I laugh at all the natural methods hitherto cry'd up; and provok'd in my own defence I venture to affirm, that not a single class before given, in any system, is natural; so long as those Genera, and those Characters, which at present exist, are arranged under it. It is easy to distribute the greatest part of the known genera into their natural classes, but so much the greater difficulty attends the arrangement of the rest. And I can not persuade myself that the present age will see a natural System, nor perhaps our latest posterity. Let us nevertheless study plants, and in the mean time content ourselves with artificial and succedaneous Classes....

10. These natural Genera assumed, (6, 7) two things are required to preserve them pure, first that the true species, and no others, be reduced to their proper genera. Secondly that all the Genera be circumscribed by true limits or boundaries, which we term *generic characters*.

11. These *characters* (10) as I turn over the authors, I find uncertain and unfix'd before *Tournefort*, to him therefore we

ought deservedly to ascribe the honor of this discovery of ascertaining the genera, indeed other Systematists of other Sects have given characters, but I can understand none of them, who preceeded *Tournefort*, or who did not tread in his steps.... *Tournefort* assumed the petals and the fruit, as diagnostic marks of the genera, and no other parts; so did almost all his followers; but the moderns, oppress'd by the quantity of new and lately detected Genera, have supposed that these parts alone are insufficient for distinguishing all the Genera: and have thence believed themselves necessitated to have recourse to the habit and appearance of plants, as the leaves, situation of the flowers, stem, root, &c. that is to recede from the steady foundation of the fructification, and to relaps into the former barbarism; with what ill omen this is done, it would be easy to demonstrate, if this were the time and place for such a talk; whatever may be the event, I acknowledge the parts described by *Tournefort* to be insufficient, if the petals alone or the fruit were to be used for this purpose. But I ask for what reason should these characters alone be used? does inspection shew this? or revelation? or any arguments either a priori, or posteriori? certainly none of these. I acknowledge no authority but inspection alone in Botany; are there not many more parts of the fructification? why should those only be acknowledged and no others? Did not the same Creator make the latter, as well as the former? are not all the parts equally necessary to the fructification? We have described of the CALYX: 1, the *Involucre*; 2, the *Spathe*; 3, the *Perianth*; 4, the *Ament*; 5, the *Glume*; 6, the *Calyptre* of the COROL; 7, the *Tube* or Claws; 8, the *Border*; 9, the *Nectary* of the STAMENS; 10, the *Filaments*; 11, the *Anthers* of the PISTIL; 12, the *Germ*; 13, the *Style*; 14, the *Stigma* of the PERICARP; 15, the *Capsule*; 16, the *Silique*; 17, the *Legume*; 18, the *Nut*; 19, the *Drupe*; 20, the *Berry*; 21, the *Pome* the SEED; 22, 23, and its *crown* the RECEPTACLE; 24, of the *fructification*; 25, of the *flower*; 26, of the *fruit*. Thus are there more parts, more letters here, than in the alphabets of languages. These marks are to us as so many vegetable letters; which, if we can read, will teach us the characters (10) of plants: they are written by the hand of God; it should be our study to read them.

12. *Tournefort* did wonders with his characters, but since so many and such new genera have been since discovered, it should be our business to adhere indeed to his principles, but to augment them with new discoveries, as the science increases.

13. *Figures* alone for determining the genera I do not recommend; before the use of letters was known to mankind, it was necessary to express everything by picture, where it could not be done by word of mouth, but on the discovery of letters the more easy and certain way of communicating ideas by writing succeeded....

a. From a figure alone who could ever argue with any certainty? but most easily from written words.

b. If I wish to bring into use, or quote in any work the characters of a Genus; I can not always easily paint the figure of it, nor etch or ingrave it, or print it off, but can easily copy the description.

c. If in the same Genus, as happens in many, some of the parts should differ in respect to number or figure in some of the species or individuals, yet I am expected to note the situation and proportion of these parts; it becomes impossible to express these by a print, unless I should give a number of figures. Hence if there should be fifty species, I must exhibit as many prints, and who would be able to extract any certainty from such a variety? But in a description the parts are omitted, which differ; and the labour is much less to describe, those which agree, and much more easy to be understood.

14. We have therefore endeavour'd to express by words all the marks or distinctions as clearly, if not more so, than others have done by their expensive prints....

19. I have selected the *marks* in describing every part of the fructification, which are certain and real, not those which are vague and fluctuating. Others have frequently assumed the taste, the odour, the size (without the proportion); these you will never find adduced by me, but only these four certain immutable mechanic principles: *Number*, *Figure*, *Situation*, and *Proportion*. These four attributes, with those twenty-six letters, above-mentioned (11), distinguish the genera so certainly from each other, that nothing more is wanted. There are other marks

for distinguishing genera, but these alone being consider'd, the rest become superfluous, nor is there any necessity to fly to the habit of the plant....

22. I am confident the *Flower* is much to be prefer'd to the *Fruit* in determining the genera, though others have been of a quite different opinion; and that the *Nectaries* are of greater advantage in determining the Genera, than almost any other part; altho' so much neglected and overlook'd by others, that they even had not a name given to them....

29. The use of some Botanic System I need not recommend even to beginners, since without system there can be no certainty in Botany. Let two enquirers, one a Systematic, and the other an Empiric, enter a garden fill'd with exotic and unknown plants, and at the same time furnish'd with the best Botanic Library; the former will easily reduce the plants by studying the letters (11) inscribed on the fructification, to their Class, Order, and Genus; after which there remains but to distinguish a few species. The latter will be necessitated to turn over all the books. to read all the descriptions, to inspect all the figures with infinite labour; nor unless by accident can be certain of his plant.

Linneus divides plants into classes according to the number and grouping of the stamens, and subdivides the classes into orders according to the number of pistils. There follow his descriptions of two genera.

55. CROCUS. (Class Triandria; Order Triandria monogynia.)

Cal. *Spathe* one-leaved.

Cor. *Tube* simple, long. *Border* six-parted, erect: the divisions egg-oblong, equal.

Stam. *Filaments* three, awl'd, shorter than the corol. *Anthers* arrow'd.

Pist. *Germ* beneath, roundish. *Style* thread-form, the length of the stamens. *Stigmas* three, convolute, saw'd.

Per. *Capsule* roundish, three-lobed, three-celled, three-valved.

Seeds numerous, round.

699. RANUNCULUS. (Class Polyandria; Order Polyandria Polygynia.)

Cal. *Perianth* five-leaved; the *leaflets* egg'd, concave, rather coloured, deciduous.

Cor. *Petals* five, obtuse, glossy: with small claws.

Nectary is a pit in each petal above the claw.

Stam. *Filaments* very numerous, shorter by half than the corol. *Anthers* erect, oblong, obtuse, twin.

Pist. *Germs* numerous, collected in a head. *Styles* none. *Stigmas* reflected, very small.

Per. none. *Receptacle* annexing the seeds with very minute peduncles.

Seeds very numerous, irregular, uncertain in figure, naked, with a reflected top.

NEW THEORIES OF EVOLUTION: LAMARCK

THE piling up of facts and their assorting brought the need for a fresh philosophy of evolution. Among several writers, perhaps the most interesting was Jean Baptiste Pierre Antoine de Monet, Chevalier de Lamarck, who was born in Picardy in 1744, the eleventh child of a poor landowner. At the age of 17, he ran away from the Jesuits' college at Amiens to take part in the Dutch wars, but soon had to leave the army owing to ill health. He studied medicine in Paris, supporting himself by working in a bank, and won fame by his botanical writings. Appointed to a zoological professorship, he extended his studies to animals, and although blind in his later years, published important works on the classification of Invertebrata, etc. He died in poverty in 1829. His claim to attention here rests on his recognition of the mutual relations and variability of species—in common with other thinkers of his age—and on his theory of the formation and modification of organs by habit—a gallant and plausible attempt to explain how evolution might have occurred. Many have since tried to reproduce such modifications experimentally, but a proof of the inheritance of acquired characters is still lacking.

PHILOSOPHIE ZOOLOGIQUE. LAMARCK

Paris: 1809

(Translation taken from *Zoological Philosophy*, by Hugh S. Elliot, 1914.)

CHAP. III. *Of species among living bodies.*

IT is not a futile purpose to decide definitely what we mean by the so-called *species* among living bodies, and to inquire if it is true that species are of absolute constancy, as old as nature, and have all existed from the beginning just as we see them to-day; or if, as a result of changes in their environment, albeit extremely slow, they have not in course of time changed their characters and shape....

Let us first see what is meant by the name of species.

Any collection of like individuals which were produced by others similar to themselves is called a species.

This definition is exact; for every individual possessing life always resembles very closely those from which it sprang; but to this definition is added the allegation that the individuals composing a species never vary in their specific characters, and consequently that species have an absolute constancy in nature.

It is just this allegation that I propose to attack, since clear proofs drawn from observation show that it is ill-founded.

The almost universally received belief is that living bodies constitute species distinguished from one another by unchangeable characteristics, and that the existence of these species is as old as nature herself. This belief became established at a time when no sufficient observations had been taken, and when natural science was still almost negligible. It is continually being discredited for those who have seen much, who have long watched nature, and who have consulted with profit the rich collections of our museums.

Moreover, all those who are much occupied with the study of natural history, know that naturalists now find it extremely difficult to decide what objects should be regarded as species.

They are in fact not aware that species have really only a constancy relative to the duration of the conditions in which are placed the individuals composing it; nor that some of these individuals have varied, and constitute races which shade gradually into some other neighbouring species....

I do not mean that existing animals form a very simple series, regularly graded throughout; but I do mean that they form a branching series, irregularly graded and free from discontinuity, or at least once free from it. For it is alleged that there is now occasional discontinuity, owing to some species having been lost. It follows that the species terminating each branch of the general series are connected on one side at least with other neighbouring species which merge into them. This I am now able to prove by means of well-known facts.

I require no hypothesis or supposition; I call all observing naturalists to witness.

Not only many genera but entire orders, and sometimes even classes, furnish instances of almost complete portions of the series which I have just indicated.

When in these cases the species have been arranged in series, and are all properly placed according to their natural affinities, if you choose one, and then, jumping over several others, take another a little way off, these two species when compared will exhibit great differences. It is thus in the first instance that we began to see such of nature's productions as lay nearest to us. Generic and specific differences were then quite easy to establish; but now that our collections are very rich, if you follow the above-mentioned series from the first species chosen to the second, which is very different from it, you reach it by slow gradations without having observed any noticeable distinctions.

I ask, where is the experienced zoologist or botanist who is not convinced of the truth of what I state?...

To assist us to a judgment as to whether the idea of species has any real foundation, let us revert to the principles already set forth; they show

(1) That all the organised bodies of our earth are true productions of nature, wrought successively throughout long periods of time.

(2) That in her procedure, nature began and still begins by fashioning the simplest of organised bodies, and that it is these alone which she fashions immediately, that is to say, only the rudiments of organisation indicated in the term *spontaneous generation.*

(3) That, since the rudiments of the animal and plant were fashioned in suitable places and conditions, the properties of a commencing life and established organic movement necessarily caused a gradual development of the organs, and in course of time produced diversity in them as in the limbs.

(4) That the property of growth is inherent in every part of the organised body, from the earliest manifestations of life; and then gave rise to different kinds of multiplication and reproduction, so that the increase of complexity of organisation, and of the shape and variety of the parts, has been preserved.

(5) That with the help of time, of conditions that necessarily were favourable, of the changes successively undergone by every part of the earth's surface, and, finally, of the power of new conditions and habits to modify the organs of living bodies, all those which now exist have imperceptibly been fashioned such as we see them.

(6) That, finally, in this state of affairs every living body underwent greater or smaller changes in its organisation and parts; so that what we call species were imperceptibly fashioned among them one after another and have only a relative constancy, and are not as old as nature....

CHAPTER VII. *Of the influence of the environment on the activities and habits of animals, and the influence of the activities and habits of these living bodies in modifying their organisation and structure.*

We are not here concerned with an argument, but with the application of a positive fact—a fact which is of more general application than is supposed, and which has not received the attention that it deserves, no doubt because it is usually very difficult to recognise. This fact consists in the influence that is exerted by the environment on the various living bodies exposed to it....

If we had not had many opportunities of clearly recognising the result of this influence on certain living bodies that we have transported into an environment altogether new and very different from that in which they were previously placed, and if we had not seen the resulting effects and alterations take place almost under our very eyes, the important fact in question would have remained for ever unknown to us.

The influence of the environment as a matter of fact is in all times and places operative on living bodies; but what makes this influence difficult to perceive is that its effects only become perceptible or recognisable (especially in animals) after a long period of time....

I must now explain what I mean by this statement: *the environment affects the shape and organisation of animals*....It is true if this statement were to be taken literally, I should be convicted of an error; for, whatever the environment may do, it does not work any direct modification whatever in the shape and organisation of animals.

But great alterations in the environment of animals lead to great alterations in their needs, and these alterations in their needs necessarily lead to others in their activities. Now if the new needs become permanent, the animals then adopt new habits which last as long as the needs that evoked them. This is easy to demonstrate, and indeed requires no amplification....

Now, if a new environment, which has become permanent for some race of animals, induces new habits in these animals, that is to say, leads them to new activities which become habitual, the result will be the use of some one part in preference to some other part, and in some cases the total disuse of some part no longer necessary.

Nothing of all this can be considered as hypothesis or private opinion; on the contrary, they are truths which, in order to be made clear, only require attention and the observation of facts.

We shall shortly see by the citation of known facts in evidence, in the first place, that new needs which establish a necessity for some part really bring about the existence of that part, as a result of efforts; and that subsequently its continued use gradually strengthens, develops and finally greatly enlarges it; in the second place, we shall see that in some cases, when the new environment and the new needs have together destroyed the utility of some part, the total disuse of that part has resulted in its gradually ceasing to share in the development of the other parts of the animal; it shrinks and wastes little by little, and ultimately, when there has been total disuse for a long period, the part in question ends by disappearing. All this is positive; I purpose to furnish the most convincing proofs of it....

Naturalists have remarked that the structure of animals is always in perfect adaptation to their functions, and have inferred that the shape and condition of their parts have determined the use of them. Now this is a mistake: for it may easily be proved by observation that it is on the contrary the needs and uses of the parts which have caused the development of these same parts, which have even given birth to them when they did not exist, and which consequently have given rise to the condition that we find in each animal.

If this were not so, nature would have had to create as many different kinds of structure in animals, as there are different kinds of environment in which they have to live; and neither structure nor environment would ever have varied.

This is indeed far from the true order of things. If things were really so, we should not have race-horses shaped like those in England; we should not have big draught-horses so heavy and so different from the former, for none such are produced in nature; in the same way we should not have basset-hounds with crooked legs, nor grey-hounds so fleet of foot, nor water-spaniels, etc.; we should not have fowls without tails, fantail pigeons, etc.; finally, we should be able to cultivate wild plants as long as we liked in the rich and fertile soil of our gardens, without the fear of seeing them change under long cultivation.

A feeling of the truth in this respect has long existed; since the following maxim has passed into a proverb and is known by all, *Habits form a second nature.*

Assuredly if the habits and nature of each animal could never vary, the proverb would have been false and would not have come into existence, nor been preserved in the event of anyone suggesting it....

The permanent disuse of an organ, arising from a change of habits, causes a gradual shrinkage and ultimately the disappearance and even extinction of that organ.

Since such a proposition could only be accepted on proof, and not on mere authority, let us endeavour to make it clear by citing the chief known facts which substantiate it.

The vertebrates, whose plan of organisation is almost the same throughout, though with much variety in their parts, have their jaws armed with teeth; some of them, however, whose environment has induced the habit of swallowing the objects they feed on without any preliminary mastication, are so affected that their teeth do not develop. The teeth then remain hidden in the bony framework of the jaws, without being able to appear outside; or indeed they actually become extinct down to their last rudiments.

In the right-whale, which was supposed to be completely destitute of teeth, M. Geoffroy has nevertheless discovered teeth concealed in the jaws of the foetus of this animal. The professor has moreover discovered in birds the groove in which the teeth should be placed, though they are no longer to be found there.

Even in the class of mammals, comprising the most perfect animals, where the vertebrate plan of organisation is carried to its highest completion, not only is the right-whale devoid of teeth, but the ant-eater (*Myrmecophaga*) is also to be found in the same condition, since it has acquired a habit of carrying out no mastication, and has long preserved this habit in its race.

Eyes in the head are characteristic of a great number of different animals, and essentially constitute a part of the plan of organisation of the vertebrates.

Yet the mole, whose habits require a very small use of sight, has only minute and hardly visible eyes, because it uses that organ so little.

Olivier's *Spalax* (*Voyage en Egypte et en Persie*), which lives underground like the mole, and is apparently exposed to daylight even less than the mole, has altogether lost the use of sight: so that it shows nothing more than vestiges of this organ. Even these vestiges are entirely hidden under the skin and other parts, which cover them up and do not leave the slightest access to light.

The *Proteus*, an aquatic reptile allied to the salamanders, and living in deep dark caves under the water, has, like the *Spalax*, only vestiges of the organ of sight, vestiges which are covered up and hidden in the same way....

It was part of the plan of organisation of the reptiles, as of other vertebrates, to have four legs in dependence on their skeleton. Snakes ought consequently to have four legs, especially since they are by no means the last order of the reptiles and are farther from the fishes than are the batrachians (frogs, salamanders, etc.).

Snakes, however, have adopted the habit of crawling on the ground and hiding in the grass; so that their body, as a result of continually repeated efforts at elongation for the purpose of passing through narrow spaces, has acquired a considerable length, quite out of proportion to its size. Now, legs would have been quite useless to these animals and consequently unused. Long legs would have interfered with their need of crawling, and very short legs would have been incapable of moving their body, since they could only have had four. The disuse of these parts thus became permanent in the various races of these animals, and resulted in the complete disappearance of these same parts, although legs really belong to the plan of organisation of the animals of this class....

The frequent use of any organ, when confirmed by habit, increases the functions of that organ, leads to its development and endows it with a size and power that it does not possess in animals which exercise it less.

We have seen that the disuse of any organ modifies, reduces and finally extinguishes it. I shall now prove that the constant use of any organ, accompanied by efforts to get the most out of it, strengthens and enlarges that organ, or creates new ones to carry on functions that have become necessary.

The bird which is drawn to the water by its need of finding there the prey on which it lives, separates the digits of its feet in trying to strike the water and move about on the surface. The skin which unites these digits at their base acquires the habit of being stretched by these continually repeated separations of the digits, thus in course of time there are formed large webs which unite the digits of ducks, geese, etc., as we actually find them. In the same way efforts to swim, that is to push against the water so as to move about in it, have stretched the membranes between the digits of frogs, sea-tortoises, the otter, beaver, etc.

On the other hand, a bird which is accustomed to perch on trees and which springs from individuals all of whom had acquired this habit, necessarily has longer digits on its feet and differently shaped from those of the aquatic animals that I have just named. Its claws in time become lengthened, sharpened and curved into hooks, to clasp the branches on which the animal so often rests.

We find in the same way that the bird of the water-side which does not like swimming and yet is in need of going to the water's edge to secure its prey, is continually liable to sink in the mud. Now this bird tries to act in such a way that its body should not be immersed in the liquid, and hence makes its best efforts to stretch and lengthen its legs. The long-established habit acquired by this bird and all its race of continually stretching and lengthening its legs, results in the individuals of this race becoming raised as though on stilts, and gradually obtaining long, bare legs, denuded of feathers up to the thighs and often higher still.

We note again that this same bird wants to fish without wetting its body, and is thus obliged to make continual efforts to lengthen its neck. Now these habitual efforts in this individual and its race must have resulted in course of time in a remarkable lengthening, as indeed we actually find in the long necks of all water-side birds.

If some swimming birds like the swan and goose have short legs and yet a very long neck, the reason is that these birds while moving about on the water acquire the habit of plunging their head as deeply as they can into it in order to get the aquatic larvae and various animals on which they feed; whereas they make no effort to lengthen their legs.

If an animal, for the satisfaction of its needs, makes repeated efforts to lengthen its tongue, it will acquire a considerable length (ant-eater, green-woodpecker); if it requires to seize anything with this same organ, its tongue will then divide and become forked. Proofs of my statement are found in the humming-birds which use their tongues for grasping things, and in lizards and snakes which use theirs to palpate and identify objects in front of them....

Nothing is more remarkable than the effect of habit in herbivorous mammals.

A quadruped, whose environment and consequent needs have for long past inculcated the habit of browsing on grass, does nothing but walk about on the ground; and for the greater part of its life is obliged to stand on its four feet, generally making only few or moderate movements. The large portion of each day that this kind of animal has to pass in filling itself with the kind of food that it cares for, has the result that it moves but little and only uses its feet for support in walking or running on the ground, and never for holding on, or climbing trees.

From this habit of continually consuming large quantities of food-material, which distend the organs receiving it, and from the habit of making only moderate movements, it has come about that the body of these animals has greatly thickened, become heavy and massive and acquired a very great size: as is seen in elephants, rhinoceroses, oxen, buffaloes, horses, etc.

The habit of standing on their four feet during the greater part of the day, for the purpose of browsing, has brought into existence a thick horn which invests the extremity of their digits; and since these digits have no exercise and are never moved and serve no other purpose than that of support like the rest of the foot, most of them have become shortened, dwindled and, finally, even disappeared.

Thus in the pachyderms, some have five digits on their feet invested in horn, and their hoof is consequently divided into five parts; others have only four, and others again not more than three; but in the ruminants, which are apparently the oldest of the mammals that are permanently confined to the ground, there are not more than two digits on the feet and indeed, in the solipeds, there is only one (horse, donkey).

Nevertheless some of these herbivorous animals, especially the ruminants, are incessantly exposed to the attacks of carnivorous animals in the desert countries that they inhabit, and they can only find safety in headlong flight. Necessity has in these cases forced them to exert themselves in swift running, and from this habit their body has become more slender and their legs much finer; instances are furnished by the antelopes, gazelles, etc.

In our own climates, there are other dangers, such as those constituted by man, with his continual pursuit of red deer, roe deer and fallow deer; this has reduced them to the same necessity, has impelled them into similar habits, and had corresponding effects.

Since ruminants can only use their feet for support, and have little strength in their jaws, which can only obtain exercise by cutting and browsing on the grass, they can only fight by blows with their heads, attacking one another with their crowns.

In the frequent fits of anger to which the males especially are subject, the efforts of their inner feeling cause the fluids to flow more strongly towards that part of their head; in some there is hence deposited a secretion of horny matter, and in others of bony matter mixed with horny matter, which gives rise to solid protuberances: thus we have the origin of horns and antlers, with which the head of most of these animals is armed.

It is interesting to observe the result of habit in the peculiar shape and size of the giraffe (*Camelo-pardalis*): this animal, the largest of the mammals, is known to live in the interior of Africa in places where the soil is nearly always arid and barren, so that it is obliged to browse on the leaves of trees and to make constant efforts to reach them. From this habit long maintained in all its race, it has resulted that the animal's fore-legs have become longer than its hind legs, and that its neck is lengthened to such a degree that the giraffe, without standing up on its hind legs, attains a height of six metres (nearly 20 feet)....

I shall show in Part II, that when the will guides an animal to any action, the organs which have to carry out that action are immediately stimulated to it by the influx of subtle fluids (the nervous fluid), which become the determining factor of the movements required. This fact is verified by many observations, and cannot now be called in question.

Hence it follows that numerous repetitions of these organised activities strengthen, stretch, develop and even create the organs necessary to them. We have only to watch attentively what is happening all around us, to be convinced that this is the true cause of organic development and changes.

Now every change that is wrought in an organ through a habit of frequently using it, is subsequently preserved by reproduction, if it is common to the individuals who unite together in fertilisation for the propagation of their species. Such a change is thus handed on to all succeeding individuals in the same environment, without their having to acquire it in the same way that it was actually created....

Everything then combines to prove my statement, namely: that it is not the shape either of the body or its parts which gives rise to the habits of animals and their mode of life; but that it is, on the contrary, the habits, mode of life, and all the other influences of the environment which have in course of time built up the shape of the body and of the parts of animals. With new shapes, new faculties have been acquired, and little by little nature has succeeded in fashioning animals such as we actually see them.

Can there be any more important conclusion in the range of natural history, or any to which more attention should be paid than that which I have just set forth?

EVOLUTION IN GEOLOGY: LYELL

LAMARCK's ingenious speculations received little credence and had little influence at the time. The next extract is from a book that deeply affected thinking men, and, in the words of Huxley, "was the chief agent in smoothing the road for Darwin"—namely *The Principles of Geology*, by Charles Lyell, published in 1830–3.

This was a final attack on the belief that a series of violent catastrophes, ending in the Flood, were responsible for the earth's present conformation. "Catastrophism," as it was called, had a history parallel with that of the doctrine of the fixity of species, raising its head in the Dark Ages, gaining the support of the Church, and flourishing especially in the eighteenth century. Its opponents, believing in the slow action of everyday forces, gradually gained strength as observation gave them new facts, and Lyell summarized the position with cogent reasoning and great literary charm.

Lyell, who lived from 1797 to 1874, belonged to a Scotch family settled in Hampshire. He became engrossed in geology at Oxford under the influence of Dr Buckland, a stout catastrophist. Travelling widely in Britain and on the continent, Lyell observed the far-reaching effects of denudation by river and sea, of volcanic eruptions and earthquakes, and began to doubt whether all the surface deposits of the earth were really the result of the Noachian deluge. His book, in three volumes, gives firstly his observations, and those of others, on the changes still proceeding on the earth, secondly his views on the distribution and origin of plant and animal species, and thirdly his explanation of the arrangement of the different layers of the earth's crust. The whole forms one continuous, closely-reasoned argument, from which we have with difficulty selected a portion to illustrate his main contention.

Lyell's book affected the progress of the theory of evolution indirectly, by disproving the interposition of the miraculous in a branch of science closely connected with the history of life. He was an evolutionist himself; in his second volume, he reasons in favour of the inter-relationship of species, but he criticises Lamarck's hypothesis as unsound, and no other had yet been suggested. After long hesitation Lyell finally came into complete agreement with Darwin concerning the theory of Natural Selection, and the two men were lifelong friends, Darwin addressing Lyell, twelve years his senior, as "My dear old Master," and subscribing himself "Your affectionate pupil."

PRINCIPLES OF GEOLOGY

BEING AN ATTEMPT TO EXPLAIN THE FORMER CHANGES OF THE EARTH'S SURFACE BY REFERENCE TO CAUSES NOW IN OPERATION

By CHARLES LYELL, Esq., F.R.S.

London: 1830–33.

VOL. III. CHAP. I.

Having considered, in the preceding volumes, the actual operation of the causes of change which affect the earth's surface and its inhabitants, we are now about to enter upon a new division of our inquiry, and shall therefore offer a few preliminary observations, to fix in the reader's mind the connexion between two distinct parts of our work, and to explain in what manner the plan pursued by us differs from that more usually followed by preceding writers on Geology.

All naturalists, who have carefully examined the arrangement of the mineral masses composing the earth's crust, and who have studied their internal structure and fossil contents, have recognised therein the signs of a great succession of former changes; and the causes of these changes have been the object of anxious inquiry. As the first theorists possessed but a scanty acquaintance with the present economy of the animate and inanimate world, and the vicissitudes to which these are subject, we find them in the situation of novices, who attempt to read a history written in a foreign language, doubting the meaning of the most ordinary terms; disputing, for example, whether a shell was really a shell,—whether sand and pebbles were the result of aqueous trituration,—whether stratification was the effect of successive deposition from water; and a thousand other elementary questions which now appear to us so easy and simple, that we can hardly conceive them to have once afforded matter for warm and tedious controversy.

In the first volume we enumerated many prepossessions which biassed the minds of the earlier inquirers, and checked an impartial desire of arriving at truth. But of all the causes to which we alluded, no one contributed so powerfully to give rise to a false method of philosophizing as the entire unconsciousness of the first geologists of the extent of their own ignorance respecting the operations of the existing agents of change.

They imagined themselves sufficiently acquainted with the mutations now in progress in the animate and inanimate world, to entitle them at once to affirm, whether the solution of certain problems in geology, could ever be derived from the observation of the actual economy of nature, and having decided that they could not, they felt themselves at liberty to indulge their imaginations, in guessing what *might be*, rather than inquiring *what is*; in other words, they employed themselves in conjecturing what might have been the course of nature at a remote period, rather than in the investigation of what was the course of nature in their own times.

It appeared to them more philosophical to speculate on the possibilities of the past, than patiently to explore the realities of the present, and having invented theories under the influence of such maxims, they were consistently unwilling to test their validity by the criterion of their accordance with the ordinary operations of nature. On the contrary, the claims of each new hypothesis to credibility appeared enhanced by the great contrast of the causes or forces introduced to those now developed in our terrestrial system during a period, as it has been termed, of *repose*.

Never was there a dogma more calculated to foster indolence and to blunt the edge of curiosity, than this assumption of the discordance between the former and the existing causes of change. It produced a state of mind unfavourable in the highest conceivable degree to the candid reception of those minute, but incessant mutations, which every part of the earth's surface is undergoing, and by which the condition of its living inhabitants is continually made to vary. The student, instead of being encouraged with the hope of interpreting the enigmas presented to him in the earth's structure,—instead of being prompted to undertake laborious inquiries into the natural history of the organic world, and the complicated effects of the igneous and aqueous causes now in operation, was taught to despond from the first. Geology, it was affirmed, could never rise to the rank of an exact science, —the greater number of phenomena must for ever remain inexplicable, or only be partially elucidated by ingenious conjectures. Even the mystery which invested the subject was said to constitute one of its principal charms, affording, as it did, full scope to the fancy to indulge in a boundless field of speculation.

The course directly opposed to these theoretical views consists in an earnest and patient endeavour to reconcile the former indications of change with the evidence of gradual mutations now in progress; restricting us, in the first instance, to known causes, and then speculating on those which may be in activity in regions inaccessible to us. It seeks an interpretation of geological monuments by comparing the changes of which they give evidence with the vicissitudes now in progress, or which may be in progress.

We shall give a few examples in illustration of the practical results already derived from the two distinct methods of theorizing, for we now have the advantage of being enabled to judge by experience of their respective merits, and by the respective value of the fruits which they have produced.

In our historical sketch of the progress of geology, the reader has seen that a controversy was maintained for more than a century, respecting the origin of fossil shells and bones,—were they organic or inorganic substances? That the latter opinion should for a long time have prevailed, and that these bodies should have been supposed to be fashioned into their present form by a plastic virtue, or some other mysterious agency, may appear absurd; but it was, perhaps, as reasonable a conjecture as could be expected from those who did not appeal, in the first instance, to the analogy of the living creation, as affording the only source of authentic information. It was only by an accurate examination of living testacea, and by a comparison of the osteology of the existing vertebrated animals with the remains found entombed in ancient strata, that this favourite dogma was exploded, and all were, at length, persuaded that these substances were exclusively of organic origin.

In like manner, when a discussion had arisen as to the nature of basalt and other mineral masses, evidently constituting a particular class of rocks, the popular opinion inclined to a belief that they were of aqueous, not of igneous origin. These rocks, it was said, might have been precipitated from an aqueous solution, from a chaotic fluid, or an ocean which rose over the continents, charged with the requisite mineral ingredients. All are now agreed that it would have been impossible for human ingenuity to invent a theory more distant from the truth; yet we must cease to wonder, on that account, that it gained so many proselytes, when we remember that its claims to probability arose partly from its confirming the assumed want of all analogy between geological causes and those now in action. By what train of investigation were all theorists brought round at length to an opposite opinion, and induced to assent to the igneous origin of these formations? By an examination of the structure of active volcanos, the mineral composition of their lavas and ejections, and by comparing the undoubted products of fire with the ancient rocks in question.

We shall conclude with one more example. When the organic origin of fossil shells had been conceded, their occurrence in strata forming some of the loftiest mountains in the world, was admitted as a proof of a great alteration of the relative level of sea and land, and doubts were then entertained whether this change might be accounted for by the partial drying-up of the ocean, or by the elevation of the solid land. The former hypothesis, although afterwards abandoned by general consent, was at first embraced by a vast majority. A multitude of ingenious speculations were hazarded to show how the level of the ocean might have been depressed, and when these theories had all failed, the inquiry, as to what vicissitudes of this nature might now be taking place, was, as usual, resorted to in the last instance. The question was agitated, whether any changes in the level of sea and land had occurred during the historical period, and, by patient research, it was soon discovered that considerable tracts of land had been permanently elevated and depressed, while the level of the ocean remained unaltered. It was therefore necessary to reverse the doctrine which had acquired so much popularity, and the unexpected solution of a problem at first regarded as so enigmatical, gave, perhaps, the strongest stimulus ever yet afforded to investigate the ordinary operations of nature. For it must have appeared almost as improbable to the earlier geologists, that the laws of earthquakes should one day throw light on the origin of mountains, as it must to the first astronomers, that the fall of an apple should assist in explaining the motions of the moon.

Of late years the points of discussion in geology have been transferred to new questions, and those, for the most part, of a higher and more general nature; but, notwithstanding the repeated warnings of experience, the ancient method of philosophising has not been materially modified.

We are now, for the most part, agreed as to what rocks are of igneous, and what of aqueous origin,—in what manner fossil shells, whether of the sea or of lakes, have been imbedded in strata,—how sand may have been converted into sandstone,—and are unanimous as to other propositions which are not of a complicated nature; but when we ascend to those of a higher order, we find as little disposition, as formerly, to make a strenuous effort, in the first instance, to search out an explanation in the ordinary economy of Nature. If, for example, we seek for the causes why mineral masses are associated together in certain groups; why they are arranged in a certain order which is never inverted; why there are many breaks in the continuity of the series; why different organic remains are found in distinct sets of strata; why there is often an abrupt passage from an assemblage of species contained in one formation to that in another immediately superimposed,—when these and other topics of an equally extensive kind are discussed, we find the habit of indulging conjectures, respecting irregular and extraordinary causes, to be still in full force.

We hear of sudden and violent revolutions of the globe, of the instantaneous elevation of mountain chains, of paroxysms of volcanic energy, declining according to some, and according to others increasing in violence, from the earliest to the latest ages. We are also told of general catastrophes and a succession of deluges, of the alternation of periods of repose and disorder, of the refrigeration of the globe, of the sudden annihilation of whole races of animals and plants, and other hypotheses, in which we see the ancient spirit of speculation revived, and a desire manifested to cut, rather than patiently to untie, the Gordian knot.

In our attempt to unravel these difficult questions, we shall adopt a different course, restricting ourselves to the known or possible operations of existing causes; feeling assured that we have not yet exhausted the resources which the study of the present course of nature may provide, and therefore that we are not authorized, in the infancy of our science, to recur to extraordinary agents. We shall adhere to this plan, not only on the grounds explained in the first volume, but because, as we have above stated, history informs us that this method has always put geologists on the road that leads to truth,—suggesting views which, although imperfect at first, have been found capable of improvement, until at last adopted by universal consent. On the other hand, the opposite method, that of speculating on a former distinct state of things, has led invariably to a multitude of contradictory systems, which have been overthrown one after the other,—which have been found quite incapable of modification,—and which are often required to be precisely reversed.

In regard to the subjects treated of in our first two volumes, if systematic treatises had been written on these topics, we should willingly have entered at once upon the description of geological monuments properly so called, referring to other authors for the elucidation of elementary and collateral questions, just as we shall appeal to the best authorities in conchology and comparative anatomy, in proof of many positions which, but for the labours of naturalists devoted to these departments, would have demanded long digressions. When we find it asserted, for example, that the bones of a fossil animal at Œningen were those of man, and the fact adduced as a proof of the deluge, we are now able at once to dismiss the argument as nugatory, and to affirm the skeleton to be that of a reptile, on the authority of an able anatomist; and when we find among ancient writers the opinion of the gigantic stature of the human race in times of old, grounded on the magnitude of certain fossil teeth and bones, we are able to affirm these remains to belong to the elephant and rhinoceros, on the same authority.

But since, in our attempt to solve geological problems, we shall be called upon to refer to the operation of aqueous and igneous causes, the geographical distribution of animals and plants, the real existence of species, their successive extinction, and so forth, we were under the necessity of collecting together a variety of facts, and of entering into long trains of reasoning, which could only be accomplished in preliminary treatises.

These topics we regard as constituting the alphabet and grammar of geology; not that we expect from such studies to obtain a key to the interpretation of all geological phenomena, but because they form the groundwork from which we must rise to the contemplation of more general questions relating to the complicated results to which, in an indefinite lapse of ages, the existing causes of change may give rise.

ORGANIC CHEMISTRY: WÖHLER

Another step towards the acceptance of the theory of evolution was the discovery that the chemical compounds found in living organisms do not, as was at first thought, owe their origin to a special vital force, but that some of them at any rate can be synthesized by man. The first production in the laboratory of one of these so-called organic compounds was effected in 1828 by a young German chemist, Wöhler, who found that he could prepare urea, a substance hitherto known only as an organic product. The dividing line was thus crossed; vital force was not involved in the production of urea, nor of other compounds of this kind soon synthesized by chemists.

Friedrich Wöhler was born near Frankfort on the Main in 1800. He studied medicine at Heidelberg and chemistry at Stockholm under Berzelius, returning to Germany to be professor of chemistry at Gottingen for 46 years. In company with Justus Liebig he laid the foundations of so-called organic chemistry, or the chemistry of the carbon compounds. He isolated and studied various chemical elements, notably aluminium, worked out many new laboratory methods, and was, throughout his long life, a successful teacher of students in all branches of chemistry.

His paper on the synthesis of urea follows in its entirety. Cyanic acid, an unstable substance, now represented by the formula HCNO, was considered by Wöhler to be C_2N_2O, or twice HCNO less H_2O. The molecules of its ammonium salt change immediately they are formed into those of urea, by a rearrangement of the component atoms.

$$(NH_4 . CNO \rightarrow NH_2 . CO . NH_2)$$

ON THE ARTIFICIAL PRODUCTION OF UREA

By F. Wöhler

Annalen der Physik und Chemie, 88, Leipzig, 1828.

In a brief earlier communication, printed in Volume III of these Annals, I stated that by the action of cyanogen on liquid ammonia, besides several other products, there are formed oxalic acid and a crystallizable white substance, which is certainly not ammonium cyanate, but which one always obtains when one attempts to make ammonium cyanate by combining cyanic acid with ammonia, e.g., by so-called double decomposition. The fact that in the union of these substances they appear to change their nature, and give rise to a new body, drew my attention anew to this subject, and research gave the unexpected result that by the combination of cyanic acid with ammonia, urea is formed, a fact that is noteworthy since it furnishes an example of the artificial production of an organic, indeed a so-called animal substance, from inorganic materials.

I have already stated that the above-mentioned white crystalline substance is best obtained by breaking down silver cyanate with ammonium chloride solution, or lead cyanate with liquid ammonia. In the latter way I prepared for myself the not unimportant amounts employed in this research. It was precipitated in colourless, transparent crystals, often more than an inch long....

With caustic soda or chalk this substance developed no trace of ammonia; with acids it showed none of the breakdown phenomena of cyanates which occur so easily, namely, the evolution of carbon dioxide and cyanic acid; neither could the lead and silver salts be precipitated from it, as from a true cyanate; it could thus contain neither cyanic acid nor ammonia as such. Since I found that by the last named method of preparation no other product was formed and the lead oxide was separated in a pure form, I imagined that an organic substance might arise by the union of cyanic acid with ammonia, possibly a substance like a vegetable salifiable base. I therefore made some experiments from this point of view on the behaviour of the crystalline substance to acids. But it was indifferent to them, nitric acid excepted; this, when added to a concentrated solution of the substance, produced at once a precipitate of glistening scales. After these had been purified by several recrystallizations, they showed very acid characters, and I was already inclined to take the compound for a real acid, when I found that after neutralization with bases it gave salts of nitric acid, from which the crystallizable substance could be extracted again with alcohol, with all the characters it had before the addition of nitric acid. This similarity to urea in behaviour induced me to make parallel experiments with perfectly pure urea separated from urine, from which I drew the conclusion that without doubt urea and this crystalline substance, or ammonium cyanate, if one can so call it, are absolutely identical compounds.

I will describe the behaviour of this artificial urea no further, since it coincides perfectly with that of urea from urine, according to the accounts of Proust, Prout and others, to be found in their writings, and I will mention only the fact, not specified by them, that both natural and artificial urea, on distillation, evolve first large amounts of ammonium carbonate, and then give off to a remarkable extent the stinging, acetic-acid-like smell of cyanic acid, exactly as I found in the distillation of mercuric cyanate or uric acid, and especially of the mercury salt of uric acid. In the distillation of urea, another white, apparently distinct substance also appears, with the examination of which I am still occupied.

But if the combination of cyanic acid and ammonia actually gives urea, it must have exactly the composition allotted to ammonium cyanate by calculation from my composition formula for the cyanates; and this is in fact the case if one atom of water is added to ammonium cyanate, as all ammonium salts contain water, and if Prout's analysis of urea is taken as the most correct. According to him, urea consists of

Nitrogen	46·650	4 atoms
Carbon	19·975	2 atoms
Hydrogen	6·670	8 atoms
Oxygen	26·650	2 atoms
	99·875	

But ammonium cyanate would consist of 56·92 cyanic acid, 28·14 ammonia, and 14·75 water, which for the separate elements gives

Nitrogen	46·78	4 atoms
Carbon	20·19	2 atoms
Hydrogen	6·59	8 atoms
Oxygen	26·24	2 atoms
	99·80	

One would have been able to reckon beforehand that ammonium cyanate with 1 atom of water has the same composition as urea, without having discovered by experiment the formation of urea from cyanic acid and ammonia. By the combustion of cyanic acid with copper oxide one obtains 2 volumes of carbon dioxide and 1 volume of nitrogen, but by the combustion of ammonium cyanate one must obtain equal volumes of these gases, which proportion also holds for urea, as Prout found.

I refrain from the considerations which so naturally offer themselves as a consequence of these facts, e.g., with respect to the composition proportions of organic substances, and the similar elementary and quantitative composition of compounds of very different properties, as for example fulminic acid and cyanic acid, a liquid hydrocarbon and olefiant gas (ethylene). From further experiments on these and similar cases, a general law might be deduced.

PASTEUR AND THE QUESTION OF SPONTANEOUS GENERATION

From the earliest times, those who sought to explain the history of life were wont to fly for help to the theory that certain forms can be spontaneously produced—from the earth, or from decaying animal and vegetable matter. There were many controversies over this theory, but light was thrown on the subject by the experiments of a French chemist, Louis Pasteur (1823–95). The son of a tanner who won the Légion d'Honneur under Napoleon, Pasteur studied at the local college and in Paris, and after occupying the chairs of chemistry in Strasbourg and then in Lille, he was appointed to a post in Paris. His first work was in the tradition of Wöhler, being concerned with the arrangement of atoms in the molecules of certain organic compounds. He next studied fermentations, showing that they were all phenomena connected with living organisms, and that when living organisms are carefully excluded, so-called spontaneous generation does not occur. We give extracts from his paper on this subject. His work at this stage was of the greatest assistance to brewers, wine and vinegar makers and dairy workers, enabling them to control their processes by scientific knowledge instead of rule of thumb methods. Applied to surgery by Sir Joseph Lister, Pasteur's discoveries were responsible for the immense advances of surgery under antiseptic conditions. He was then asked to investigate silkworm disease, and afterwards studied chicken cholera, anthrax and rabies, showing each disease to be due to a specific parasite which he isolated and grew on suitable media in his laboratory. In each case he found a method of cure or prevention.

The success of Pasteur's work may be ascribed to the patience and insight with which he applied exact chemical and physical methods to the complex problems of fermentation and disease. He died at the age of 73 with his life's motto on his lips: "il faut travailler."

MEMOIR *on* THE ORGANIZED CORPUSCLES WHICH EXIST IN THE ATMOSPHERE

AN EXAMINATION *of the* DOCTRINE OF SPONTANEOUS GENERATION

By M. L. PASTEUR

Annales de Chimie et de Physique (1862), LXIV, p. 5.

Chapter I. HISTORICAL.

IN ancient times, and until the end of the middle ages, everyone believed in the occurrence of spontaneous generations. Aristotle says that animals are engendered by all dry things that become moist and all moist things that become dry.

Van Helmont describes the way to bring mice into existence.

Even in the seventeenth century, many authors give methods for producing frogs from the mud of marshes, or eels from river water.

Such errors could not survive for long the spirit of investigation which arose in Europe in the sixteenth and seventeenth centuries.

Redi, a celebrated member of the *Academia del Cimento*, demonstrated that the worms in putrefying flesh were larvae from the eggs of flies. His proofs were as simple as they were decisive, for he showed that surrounding the putrefying flesh with fine gauze absolutely prevented the appearance of these larvae....

But, in the second part of the seventeenth and the first part of the eighteenth centuries, microscopic observations rapidly increased in number. The doctrine of spontaneous generation then reappeared. Some, unable to explain the origin of the varied organisms which the microscope showed in their infusions of animal or vegetable matters, and seeing nothing among them which resembled sexual reproduction, were obliged to assume that matter which has once lived keeps, after its death, a special vital force, under the influence of which its scattered particles unite themselves afresh under certain favourable conditions with varieties of structure determined by these conditions.

Others, on the contrary, used their imagination to extend the marvellous revelations of the microscope, and believed they saw males, females, and eggs among these Infusoria, and they consequently set themselves up as open adversaries of spontaneous generation.

One must recognize that the proofs in support of either of these opinions scarcely bore examination....[Pasteur describes later experiments at length.]

After the work of which I have spoken, the Académie des Sciences, realizing how much still remained to be found out, offered a prize for a dissertation on the following subject: "*An endeavour, by accurate experiments, to throw light on the question of spontaneous generation.*"

The problem then appeared so obscure that M. Biot, whose kindness with regard to my work has always been unfailing, expressed his regret at seeing me engaged on these researches, and claimed from me a promise to abandon the subject after a limited time if I had not overcome the difficulties which were then perplexing me. M. Dumas, who has often conspired with M. Biot in kindness to me, said to me about the same time: "I should not advise anyone to spend too long over this subject."

What need had I to concern myself with it?

Chemists had discovered, twenty years earlier, a collection of extraordinary phenomena comprised under the generic term "*fermentations.*" Two classes of substance are concerned in them all: one known as *fermentable*, such as sugar; the other *nitrogenous*, always an albumen-like substance. This was the theory which was universally accepted: albuminous substances undergo a change on exposure to the air, a special oxidation of unknown nature, which gives them the character of a *ferment*, that is to say, the property of subsequently acting, by their contact, on fermentable substances.

There was certainly one ferment, the oldest and most remarkable of all, which was known to be a living organism:—brewer's yeast. But in all fermentations discovered more recently than the organic nature of yeast (1836), the existence of living organisms had not been recognized even after careful examination. Physiologists had therefore gradually abandoned, though with regret, M. Cagniard de Latour's hypothesis of a probable relation between the living nature of yeast and its property of being a ferment. The general theory was applied to yeast in such terms as these:—"Yeast is not active because it is an organism, but because it has been in contact with the air. It is the dead yeast cells, those which have lived and are in the process of decay, which act upon the sugar."

My studies led me to entirely different conclusions. I found that all fermentations properly so called, lactic, butyric, the fermentations of tartaric acid, of malic acid, of urea, were always connected with the presence and multiplication of living organisms. Far from the living nature of yeast being an obstacle to the theory of fermentation, it was that very fact which made it also subject to the common law and established it as the type of all true ferments. My conclusion was that albuminous substances are never ferments, but the food of ferments. True ferments are living organisms.

Accepting this, it was said that ferments originated from the contact of albuminous substances with oxygen gas. One of two things must be true, I said to myself: ferments are organized; if they are produced by oxygen alone, considered merely as oxygen, they are spontaneously generated; if they are not of spontaneous origin, it is not as oxygen alone that this gas intervenes in their formation, but as a stimulant to germs entering with it, or existing in the nitrogenous or fermentable matters. At this point in my study of fermentations, I had thus to form an opinion on the question of spontaneous generation. I might perhaps find therein a powerful weapon in aid of my ideas on true fermentations.

The researches which I am about to describe were consequently a necessary digression from my work on fermentations....

Chapter II. The Microscopic examination of the solid particles found in the atmosphere.

My first care was to find a method which should permit of collecting the solid particles that float in the air and of studying them under the microscope. It was first necessary to remove if possible the objections held by the partisans of spontaneous generation against the ancient hypothesis of the aerial dissemination of germs....

The method which I used to collect and examine the dusts suspended in the air is very simple; it consists in filtering a known volume of air through gun-cotton, which is soluble in a mixture of alcohol and ether. The solid particles collect on the fibres of the cotton. The cotton is then treated with its solvent, and after a time all the solid particles fall to the bottom of the liquid; they are washed several times and transferred to the stage of the microscope, where they are easily examined....

These simple manipulations allow one to observe that ordinary air always contains a variable number of corpuscles whose form and structure show them to be of organic nature. In size they vary from the smallest diameters up to $\frac{1}{100}$ to $\frac{1.5}{100}$ of a millimetre, or more. Some are perfectly spherical, others oval. Their outlines are more or less sharply defined. Some are quite transparent, but others have a granular substance and are opaque. Those which are transparent with clearly defined outlines are so much like the spores of common moulds that the cleverest microscopist could not distinguish between them. That is all that one can say, just as one can only affirm that among the others there are some which resemble encysted Infusoria or the globules which are regarded as the eggs of these minute creatures....

I found that a little wad of cotton, thus exposed for twenty-four hours in the summer after a spell of fine weather to a current of one litre of air a minute from the Rue d'Ulm, collects several thousands of organized corpuscles. The number varies indefinitely with the state of the atmosphere—before or after rain, in still or windy weather, by day or by night, near the ground or at some distance from it....

Chapter III. Experiments with Calcined Air.

As we have just seen, organized corpuscles are always to be found suspended in the air; these cannot be distinguished from the germs of inferior organisms by their shape, size, or apparent structure, and are present in numbers that, without exaggeration, are indeed great. Do fertile germs really exist among them? Obviously this was the interesting question; I believe I have found a definite answer. But before describing the experiments which bear more particularly on this side of the subject, it is necessary to consider whether Dr Schwann's observations on the inactivity of air which has been heated are well established....

Into a flask of 250 to 300 cubic centimetres capacity I introduce 100 to 150 cubic centimetres of a fluid of the following composition:

water	100
sugar	10
albuminous and mineral matters from brewer's yeast	0·2 to 0·7

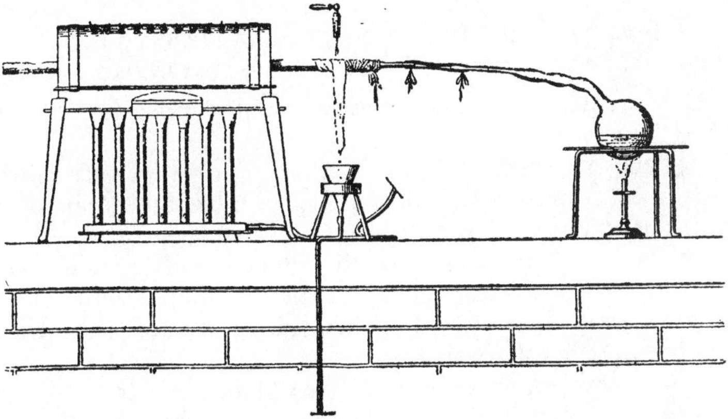

Fig. 1.

The drawn out neck of the flask communicates with a tube of platinum kept at red heat (see Fig. 1). The liquid is boiled for two or three minutes and then allowed to cool. The flask fills slowly with ordinary air at atmospheric pressure, all of which has been heated; the neck of the flask is then closed in the flame (Fig. 2).

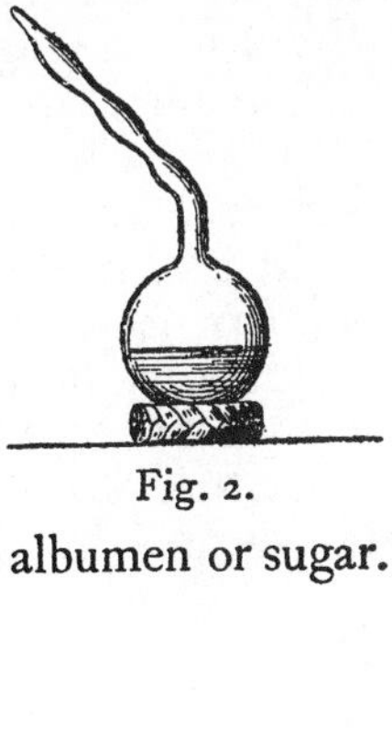

Fig. 2.

The flask thus prepared is put in an incubator at a constant temperature of about 30°; the liquid keeps indefinitely without the slightest alteration. Its limpidity, its odour, its feebly acid reaction, show no appreciable change. Its colour deepens slowly with time, doubtless under the influence of a direct oxidation of the albumen or sugar.

I affirm, with the utmost sincerity, that I have never had a doubtful result from a single experiment of this kind. Sugared yeast-water boiled for two or three minutes and then exposed to air which has been heated never alters at all, even after eighteen months at a temperature of 25° to 30°, while if one abandons it to ordinary air, after a day or two it is seen to be in the course of a manifest change and becomes full of bacteria and vibrios, or covered with moulds.

Dr Schwann's observation is thus fully confirmed when applied to sugared yeast-water....

Chapter IV. SOWING OF DUSTS FROM AIR INTO LIQUIDS SUITABLE FOR THE DEVELOPMENT OF INFERIOR ORGANISMS.

The results of the experiments of the two preceding chapters have taught us:

1. that in suspension in ordinary air there are always organized corpuscles which closely resemble the germs of inferior organisms;

2. that sugared yeast-water, though eminently alterable when exposed to ordinary air, remains unaltered, limpid, without producing Infusoria or moulds, when left in contact with air that has been previously heated.

This admitted, let us endeavour to find out what will happen if into this albumen-containing sugar solution are sown the dusts of which the collection is described in chapter II, without introducing anything else but the dusts, and only air that has been heated....

I adopted the following procedure to introduce dusts from air into putrefiable or fermentable liquids, in presence of heated air.

Let us take again our flask containing sugared yeast-water and air that has been heated. I will suppose that the flask has been in the incubator at 25° or 30° for one or two months, without having shown a noticeable alteration—a manifest proof of the inactivity of the heated air with which it was filled under ordinary atmospheric pressure.

The neck of the flask remaining closed, it is joined by means of a rubber tube to an apparatus arranged as shown in Fig. 3: *T* is a tube of hard glass, of 10–12 millimetres interior diameter, in which I have placed a scrap of glass tubing *a* of narrow diameter, open at the ends, free to move in the large tube, and containing a portion of one of the little wads of cotton loaded with dust: *R* is a brass tube of *T* shape fitted with taps, of which one communicates with the air-pump, another with a platinum tube at red heat, the third with the tube *T*; *cc* represents the rubber which joins the flask *B* to the tube *T*.

When all the parts of the apparatus are arranged and when the platinum tube has been brought to red heat by the gas burner shown at *G*, the tap which leads to the platinum tube is

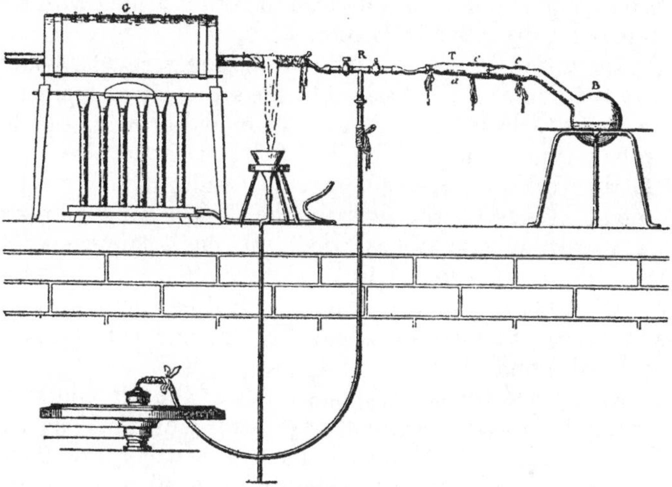

Fig. 3.

closed, and the whole is evacuated by means of the pump. The tap is then opened so as to allow the calcined air to enter the apparatus slowly. The evacuation and the re-entry of the calcined air are repeated alternately ten or twelve times. The little tube containing the cotton is thus filled even to the smallest interstices of the cotton with air that has been heated, but the dusts remain in it. The tip of the flask *B* is then broken off through the rubber without undoing the fastenings, and the little tube with the dusts is made to slide into the flask. Finally, the neck of the flask is closed in the flame and it is replaced in the incubator. Now, it always happens that growths begin to appear after 24, 36, or 48 hours at most.

This is precisely the time necessary for these same growths to appear in sugared yeast-water when it is exposed to ordinary air.

Here are the details of (one) experiment.

Early in November, 1859, I prepared several flasks of 250 cubic centimetres capacity, containing 100 cubic centimetres of sugared yeast-water and 150 cubic centimetres of heated air. They remained in the incubator at a temperature of about 30° till the 8th of January, 1860. On that day, about 9 a.m., I introduced into one of these flasks, with the help of the apparatus shown in Fig. *c*, a portion of a wad of cotton loaded with dusts, collected as I described in chapter II.

On the 9th of January, at 9 a.m., nothing particular could be seen in the liquid in the flask. On the same day at 6 p.m., one could see very distinctly little tufts of mould growing out from the tube with the dusts. The liquid was perfectly clear.

On the 10th of January at 5 p.m., besides the silky tufts of the mould, I saw on the walls of the flask a large number of white streaks which looked iridescent on holding the flask between the eye and the light.

On the 11th of January, the liquid had lost its clearness. It was all turbid, to such an extent that one could no longer see the tufts of mould.

I then opened the flask by a scratch of the file and studied under the microscope the different growths which had appeared.

The turbidity of the liquid was due to a crowd of little bacteria of the smallest dimensions, very rapid in their movements, spinning sharply or swaying to and fro, etc. (Fig. 4).

The silky tufts were formed by a mycelium of branching tubes (Fig. 5).

Finally, the precipitate which showed itself on the 10th of January in the form of white streaks was composed of a very delicate Torula (Fig. 6)...resembling brewer's yeast, but with smaller cells....

Here then were three growths derived from the dusts which had been added, growths of the same kind as those which appear in similar sugared albuminous liquids when they are abandoned to the contact of ordinary air....

I could multiply many times such examples of growths in sugared yeast-water following on the addition of dusts from air, in an atmosphere of air previously heated and of itself quite inactive. I have chosen for description an experiment which showed very common organisms, which occur frequently in such liquids as those which I employed. But the most diverse organisms may appear....

One might perhaps wonder if, in the preceding experiments, the cotton, as an organic substance, had some influence on the results. It would above all be useful to know what would happen if similar manipulations were carried out on flasks prepared as

Fig. 4.

Fig. 5.

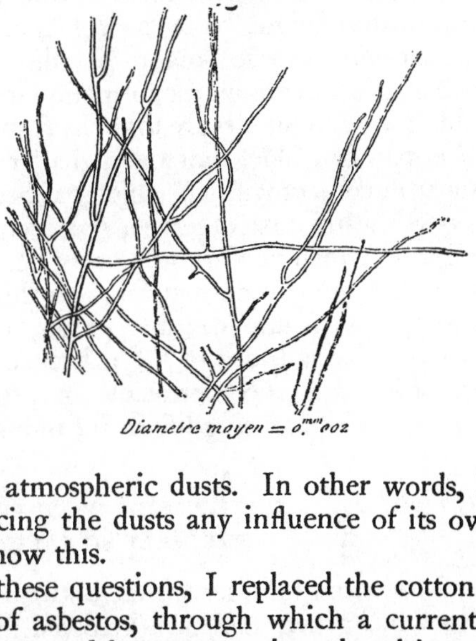

Fig. 6.

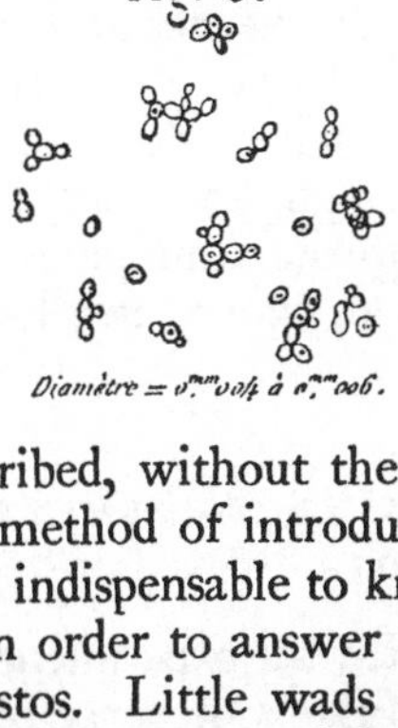

Diamètre = 0mm 004 à 0mm 006.

Diametre moyen = 0mm 002

described, without the atmospheric dusts. In other words, has the method of introducing the dusts any influence of its own? It is indispensable to know this.

In order to answer these questions, I replaced the cotton by asbestos. Little wads of asbestos, through which a current of air had been passed for several hours, were introduced into the flasks according to the preceding instructions, and they gave results of exactly the same kind as those we have just quoted. But with wads of asbestos previously calcined and not filled with dust, or filled with dust but heated afterwards, no turbidity, nor Infusoria, nor plants of any kind were ever produced. The liquids remained perfectly clear. I have repeated these comparative experiments very many times, and I have always been surprized by their distinctness, by their perfect constancy. It would seem, indeed, that experiments of this delicacy should sometimes show contradictory results due to accidental causes of error. But never once did any of my blank experiments show any growths, just as the sowing of dusts has always furnished living organisms.

In face of such results, confirmed and enlarged by those of the following chapters, I regard this as mathematically demonstrated: all organisms which appear in sugared albuminous solutions boiled and then exposed to ordinary air derive their origin from the solid particles which are suspended in the atmosphere.

But, on the other hand, we have seen in chapter II that these solid particles include, amid a multitude of amorphous fragments —carbonate of lime, silica, soot, bits of wool, etc.,—organized corpuscles which are so like as to be indistinguishable from the little spores of the growths whose formation we have recognized in this liquid. These corpuscles are then the fertile germs of the growths.

We may conclude, moreover, that if an albuminous solution of sugar in contact with air which has been heated does not alter, as Dr Schwann first observed, it is because the heat destroyed the germs which the air was carrying. All the adversaries of spontaneous generation had foreseen this. I have done nothing but supply sure and decisive proofs, obliging non-prejudiced minds to reject all idea of the existence in the atmosphere of a more or less mysterious principle, gas, fluid, ozone, etc., having the property of arousing life in infusions....

Chapter IX. On the mode of nutrition of ferments properly so called.

It is essential to observe that until to-day all experiments on spontaneous generation have involved infusions of vegetable or animal matters, that is, liquids containing substances which have previously been part of some organism. Whatever may be the conditions of temperature to which one exposes them, these substances have a constitution and properties acquired under the influence of life.

This fact has served as a groundwork for all theories of spontaneous generation. Now, I am going to demonstrate in this chapter that the appearance of inferior organisms does not necessarily presuppose the presence of plastic organic substances, of those albuminous substances which the chemist has never been able to produce, and which for their formation require the aid of vital forces....

Experiment has shown me, in fact, that in the experiments of chapters IV, V, VI, one can replace sugared yeast-water, urine, milk, etc., by a fluid of the following composition:

pure water	100
crystalline sugar	10
tartrate of ammonia	0·2 to 0·5
ash of brewer's yeast	0·1

If one sows into this liquid, in presence of calcined air only, the dusts which exist suspended in air, one finds appearing therein bacteria, vibrios, moulds, etc. The albumens, the fats, the essential oils, the colouring matters proper to these organisms, are formed in all cases from the elements of ammonia, phosphates, and sugar.

The fluid may be composed in the same way with the addition of chalk:

pure water	100
crystalline sugar	10
tartrate of ammonia	0·2 to 0·5
ash of brewer's yeast	0·1
pure chalk	3 to 5

and the same phenomena take place, but with a more marked tendency towards the fermentations known as lactic and butyric, and the vegetable or animal ferments proper to these fermentations will arise simultaneously or successively.

I shall publish shortly a detailed work on the results I obtained in these studies, which have always seemed to me to offer much of great interest in connection with so-called spontaneous generation.

It was from these results that I was led to undertake the following experiments, whose success surpassed my expectations.

In pure distilled water I dissolve a crystalline salt of ammonia, crystalline sugar, and phosphates from the calcination of yeast; I then sow into the liquid a few spores of *Penicillium* or of some mould. These spores germinate easily, and after only two or three days, the liquid is filled with flakes of mycelium, of which a large number spread themselves on the surface of the liquid, where they fructify. The vegetation is not in the least sickly or feeble. The precaution of employing an acid salt of ammonia hinders the development of Infusoria, which would soon stop the progress of the little plant through absorbing the oxygen of the air, without which the mould cannot thrive. All the carbon of the plant is obtained from the sugar, which is slowly used up, all its nitrogen from the ammonia, its mineral matters from the phosphates. In this matter of the assimilation of nitrogen and phosphates there is thus a complete analogy between ferments, moulds, and plants of a complicated organization. The following facts complete the proof of this in a definitive manner.

If, in the experiments I have just described, I omit any one of the dissolved principles, growth is arrested. For example, the mineral matter might appear the least indispensable for organisms of this nature. But, if the liquid is deprived of phosphates, vegetation is no longer possible, whatever may be the proportions of sugar and ammonium salts. The germination of the spores just begins under the influence of the phosphates which are introduced in infinitely small amounts with the spores one has sown. In the same way, if one omits the salt of ammonia, the plant shows no development. There is only an insignificant beginning of germination, due to the presence of the albuminous matter of the spores, although there is an abundance of free nitrogen in the surrounding air and dissolved in the liquid. Finally, the same happens if one omits the sugar, the carbon-containing foodstuff, although there may be in the air or in the liquid a certain proportion of carbonic acid. All proclaims the fact that, with regard to the origin of their carbon, fungi differ essentially from phanerogamic plants. They do not decompose carbonic acid; they do not set free oxygen. The absorption of oxygen and the setting free of carbonic acid are on the contrary the necessary and permanent activities of their life.

These facts give us precise ideas on the mode of nutrition of fungi, with respect to which science does not yet possess connected observations.

They also—a fact which one should perhaps preferably emphasize—exhibit a method with the help of which vegetable physiology will be able to approach without difficulty the most intricate problems of the life of these little plants, in such a way as to prepare an assured path for the study of the same problems among higher plants.

Even if one fears the impossibility of applying to higher plants the results furnished by organisms of so lowly an appearance, there is none the less a great interest in resolving the difficulties raised by the study of the life of plants, beginning with those in which the least complicated organisation makes conclusions easier and more sure: the plant is here reduced in some sort to a cellular state, and the progress of science shows more and more that the study of reactions occurring under the influence of animal or vegetable life in their most complicated manifestations leads back after all to the elucidation of phenomena proper to the cell.

THE ORIGIN OF SPECIES

THE time was now ripe for the acceptance of the theory of evolution. The foundations of the idea of special creation had been weakened or removed one by one. Embryology had shown that animals resemble one another even more closely in the early stages of their development than when adult. Geology had suggested that natural forces still at work could account for all known structures of the earth's crust. Chemical laws were discovered to hold good within living tissues; Pasteur had put an end to the hoary myth of spontaneous generation by his accurate and beautiful experiments. Naturalists had begun to doubt the fixity of species; Lamarck and others had freely speculated on evolutionary processes.

Finally the great naturalist Charles Darwin laid his finger on a factor of evolution that can be seen at work—the natural selection of the creatures best fitted to survive in the struggle for existence. His book, aided by the arguments of his brilliant supporter Huxley, carried the position. An upheaval in popular thought took place. After much conflict and discussion, the view that evolution had occurred was widely accepted. Definite research was begun on natural selection and other factors which might be considered to be involved in the process.

Charles Darwin was the son of a successful doctor, and grandson of a zoologist of repute on his father's side and of Josiah Wedgwood the potter on his mother's. He was born at Shrewsbury in 1809, studied at Edinburgh for medicine and at Cambridge for the Church, but received his chief scientific training during the five years' voyage of H.M.S. "Beagle" on a surveying expedition to which he was appointed naturalist. His observations led him to reflect on the question of the transmutation of species. He started notebooks for the collection of facts, especially investigating the variations occurring in domestic animals, and he saw the important effect of man's selection. After considering how selection could act in nature, he read, in 1838, Malthus' Essay on Population, in which is described the struggle of all living beings for existence, consequent on their high natural rate of increase. It struck Darwin that under the circumstances of this struggle, which he had himself often observed, favourable variations would tend to be preserved, and to be perpetuated in their offspring, while unfavourable ones would be destroyed. He wrote out a sketch of this theory in 1842, but published nothing till 1859, when after 20 years of work, his book *The Origin of Species by means of Natural Selection* was given to the world. It contains the evidence for evolution itself, in a more complete form than ever before, the facts of variation in domestic animals and under nature, an account of the struggle for existence, and of natural selection as the result of the struggle acting on the variations of species. He shows this power as a sufficient explanation of the evolutionary process, and of the adaptation of creatures to their environment.

From Darwin's work we have chosen (1) the introduction of the Origin of Species, giving the plan of the book, (2) parts of Chapter IV on Natural Selection, which he calls the keystone of his arch, and (3) extracts from a later work on the Descent of Man to illustrate his marshalling of facts in support of the theory of evolution. A page from Malthus' essay is placed first, since this gave the clue to Natural Selection both to Darwin and to his contemporary Alfred Russel Wallace.

AN ESSAY ON THE PRINCIPLE OF POPULATION

By the Rev. T. R. MALTHUS, A.M., F.R.S.

(First published in 1798)

Extract from the beginning of the first chapter.

In an inquiry concerning the improvement of society, the mode of conducting the subject which naturally presents itself, is:

1. To investigate the causes that have hitherto impeded the progress of mankind towards happiness; and,

2. To examine the probability of the total or partial removal of these causes in future.

To enter fully into this question, and to enumerate all the causes that have hitherto influenced human improvement, would be much beyond the power of an individual. The principal object of the present essay is to examine the effects of one great cause intimately united with the very nature of man; which, though it has been constantly and powerfully operating since the commencement of society, has been little noticed by the writers who have treated this subject. The facts which establish the existence of this cause have, indeed, been repeatedly stated and acknowledged; but its natural and necessary effects have been almost totally overlooked; though probably among these effects may be reckoned a very considerable portion of that vice and misery, and of that unequal distribution of the bounties of nature, which it has been the unceasing object of the enlightened philanthropist in all ages to correct.

The cause to which I allude, is the constant tendency in all animated life to increase beyond the nourishment prepared for it.

It is observed by Dr Franklin, that there is no bound to the prolific nature of plants or animals but what is made by their crowding and interfering with each other's means of subsistence. Were the face of the earth, he says, vacant of other plants, it might be gradually sowed and overspread with one kind only, as, for instance, with fennel: and were it empty of other inhabitants, it might in a few ages be replenished from one nation only, as for instance, with Englishmen.

This is incontrovertibly true. Throughout the animal and vegetable kingdoms Nature has scattered the seeds of life abroad with the most profuse and liberal hand; but has been comparatively sparing in the room and the nourishment necessary to rear them. The germs of existence contained in this earth, if they could freely develop themselves, would fill millions of worlds in the course of a few thousand years. Necessity, that imperious, all-pervading law of nature, restrains them within the prescribed bounds. The race of plants and the race of animals shrink under this great restrictive law; and man cannot by any efforts of reason escape from it.

THE ORIGIN OF SPECIES

BY MEANS OF NATURAL SELECTION, OR THE PRESERVATION OF FAVOURED RACES IN THE STRUGGLE FOR LIFE

By CHARLES DARWIN, M.A., F.R.S., &c.

(London: 1859)

INTRODUCTION.

When on board H.M.S. "Beagle," as naturalist, I was much struck with certain facts in the distribution of the inhabitants of South America, and in the geological relations of the present to the past inhabitants of that continent. These facts seemed to me to throw some light on the origin of species—that mystery of mysteries, as it has been called by one of our greatest philosophers. On my return home, it occurred to me, in 1837, that something might perhaps be made out on this question by patiently accumulating and reflecting on all sorts of facts which could possibly have any bearing on it. After five years' work I allowed myself to speculate on the subject, and drew up some short notes; these I enlarged in 1844 into a sketch of the conclusions, which then seemed to me probable: from that period to the present day I have steadily pursued the same object. I hope that I may be excused for entering on these personal details, as I give them to show that I have not been hasty in coming to a decision.

My work is now nearly finished; but as it will take me two or three more years to complete it, and as my health is far from strong, I have been urged to publish this abstract. I have more especially been induced to do this, as Mr Wallace, who is now studying the natural history of the Malay Archipelago, has arrived at almost exactly the same general conclusions that I have on the origin of species. In 1858 he sent to me a memoir on this subject, with a request that I would forward it to Sir Charles Lyell, who sent it to the Linnean Society, and it is published in the third volume of the Journal of that Society. Sir C. Lyell and Dr Hooker, who both knew of my work—the latter having read my sketch of 1844—honoured me by thinking it advisable to publish, with Mr Wallace's excellent memoir, some brief extracts from my manuscripts....

In considering the Origin of Species, it is quite conceivable that a naturalist, reflecting on the mutual affinities of organic beings, on their embryological relations, their geographical distribution, geological succession, and other such facts, might come to the conclusion that species had not been independently created, but had descended, like varieties, from other species. Nevertheless, such a conclusion, even if well founded, would be unsatisfactory, until it could be shown how the innumerable species inhabiting this world have been modified, so as to acquire that perfection of structure and coadaptation which most justly excites our admiration. Naturalists continually refer to external conditions, such as climate, food, &c., as the only possible cause of variation. In one limited sense, as we shall hereafter see, this may be true; but it is preposterous to attribute to mere external conditions, the structure, for instance, of the woodpecker, with its feet, tail, beak, and tongue, so admirably adapted to catch insects under the bark of trees. In the case of the mistletoe, which draws its nourishment from certain trees, which has seeds that must be transported by certain birds, and which has flowers with separate sexes absolutely requiring the agency of certain insects to bring pollen from one flower to the other, it is equally preposterous to account for the structure of this parasite, with its relations to several distinct organic beings, by the effects of external conditions, or of habit, or of the volition of the plant itself....

It is, therefore, of the highest importance to gain a clear insight into the means of modification and coadaptation. At the commencement of my observations it seemed to me probable that a careful study of domesticated animals and of cultivated plants would offer the best chance of making out this obscure problem. Nor have I been disappointed; in this and in all other perplexing cases I have invariably found that our knowledge, imperfect though it be, of variation under domestication, afforded the best and safest clue. I may venture to express my conviction of the high value of such studies, although they have been very commonly neglected by naturalists.

From these considerations, I shall devote the first chapter of this Abstract to Variation under Domestication. We shall thus see that a large amount of hereditary modification is at least possible; and, what is equally or more important, we shall see how great is the power of man in accumulating by his Selection successive slight variations. I will then pass on to the variability of species in a state of nature; but I shall, unfortunately, be compelled to treat this subject far too briefly, as it can be treated properly only by giving long catalogues of facts. We shall, however, be enabled to discuss what circumstances are most favourable to variation. In the next chapter the Struggle for Existence amongst all organic beings throughout the world, which inevitably follows from their high geometrical powers of increase, will be treated of. This is the doctrine of Malthus, applied to the whole animal and vegetable kingdoms. As many more individuals of each species are born than can possibly survive; and as, consequently, there is a frequently recurring struggle for existence, it follows that any being, if it vary however slightly in any manner profitable to itself, under the complex and sometimes varying conditions of life, will have a better chance of surviving, and thus be *naturally selected.* From the strong principle of inheritance, any selected variety will tend to propagate its new and modified form.

This fundamental subject of Natural Selection will be treated at some length in the fourth chapter; and we shall then see how Natural Selection almost inevitably causes much Extinction of the less improved forms of life, and leads to what I have called Divergence of Character. In the next chapter I shall discuss the complex and little known laws of variation and of correlation of growth. In the four succeeding chapters, the most apparent and gravest difficulties in accepting the theory will be given: namely, first, the difficulties of transitions, or how a simple being or a simple organ can be changed and perfected into a highly developed being or into an elaborately constructed organ; secondly, the subject of Instinct, or the mental powers of animals; thirdly, Hybridism, or the infertility of species and the fertility of varieties when intercrossed; and fourthly, the imperfection of the geological record. In the next chapter I shall consider the geological succession of organic beings throughout time; in the eleventh and twelfth, their geographical distribution throughout space; in the thirteenth, their classification or mutual affinities, both when mature and in an embryonic condition.

In the last chapter I shall give a brief recapitulation of the whole work, and a few concluding remarks.

No one ought to feel surprise at much remaining as yet unexplained in regard to the origin of species and varieties, if he make due allowance for our profound ignorance in regard to the mutual relations of the many beings which live around us. Who can explain why one species ranges widely and is very numerous, and why another allied species has a narrow range and is rare? Yet these relations are of the highest importance, for they determine the present welfare, and, as I believe, the future success and modification of every inhabitant of this world. Still less do we know of the mutual relations of the innumerable inhabitants of the world during the many past geological epochs in its history. Although much remains obscure, and will long remain obscure, I can entertain no doubt, after the most deliberate study and dispassionate judgment of which I am capable, that the view which most naturalists entertain, and which I formerly entertained—namely, that each species has been independently created—is erroneous. I am fully convinced that species are not immutable; but that those belonging to what are called the same genera are lineal descendants of some other and generally extinct species, in the same manner as the acknowledged varieties of any one species are the descendants of that species. Furthermore, I am convinced that Natural Selection has been the main, but not exclusive, means of modification.

Chapter IV. Natural Selection; or the Survival of the Fittest.

How will the struggle for existence, discussed too briefly in the last chapter, act in regard to variation? Can the principle of selection, which we have seen is so potent in the hands of man, apply in nature? I think we shall see that it can act most effectually. Let it be borne in mind in what an endless number of strange peculiarities in our domestic productions, and, in a lesser degree, those under nature, vary; and how strong the hereditary tendency is. Under domestication, it may be truly said that the whole organisation becomes in some degree plastic. Let it be borne in mind how infinitely complex and close-fitting are the mutual relations of all organic beings to each other and to their physical conditions of life. Can it, then, be thought improbable, seeing that variations useful to man have undoubtedly occurred, that other variations useful in some way to each being in the great and complex battle of life, should sometimes occur in the course of thousands of generations? If such do occur, can we doubt (remembering that many more individuals are born than can possibly survive) that individuals having any advantage, however slight, over others, would have the best chance of surviving and of procreating their kind? On the other hand, we may feel sure that any variation in the least degree injurious would be rigidly destroyed. This preservation of favourable variations, and the rejection of injurious variations, I call Natural Selection. Variations neither useful nor injurious would not be affected by natural selection, and would be left a fluctuating element, as perhaps we see in the species called polymorphic....

As man can certainly produce great results [with his domestic animals and plants] by adding up in any given direction mere individual differences, so could Nature, but far more easily, from having incomparably longer time at her disposal....

Man can act only on external and visible characters: nature cares nothing for appearances, except in so far as they may be useful to any being. She can act on every internal organ, on every shade of constitutional difference, on the whole machinery of life. Man selects only for his own good: Nature only for that of the being which she tends....

It may be said that natural selection is daily and hourly scrutinising, throughout the world, every variation, even the slightest, rejecting that which is bad, preserving and adding up all that is good; silently and insensibly working, whenever and wherever opportunity offers at the improvement of each organic being in relation to its organic and inorganic conditions of life. We see nothing of these slow changes in progress, until the hand of time has marked the long lapse of ages, and then so imperfect is our view into long-past geological ages, that we see only that the forms of life are now different from what they formerly were.

Although natural selection can act only through and for the good of each being, yet characters and structures which we are apt to consider as of very trifling importance, may be thus acted on. When we see leaf-eating insects green, and bark-feeders mottled-grey, the alpine ptarmigan white in winter, the red-grouse the colour of heather and the black-grouse that of peaty earth, we must believe that these tints are of service to these birds and insects in preserving them from danger. Grouse, if not destroyed at some period of their lives, would increase in countless numbers; they are known to suffer largely from birds of prey; and hawks are guided by eyesight to their prey—so much so, that on parts of the Continent persons are warned not to keep white pigeons, as being the most liable to destruction. Hence I can see no reason to doubt that natural selection might be most effective in giving the proper colour to each kind of grouse, and in keeping that colour, when once acquired, true and constant. Nor ought we to think that the occasional destruction of an animal of any particular colour would produce little effect: we should remember how essential it is in a flock of white sheep to destroy every lamb with the faintest trace of black. In plants the down on the fruit and the colour of the flesh are considered by botanists as characters of the most trifling importance: yet we hear from an excellent horticulturalist, Downing, that in the United States smooth-skinned fruits suffer far more from a beetle, a Curculio, than those with down; that purple plums suffer far more from a certain disease than yellow plums, whereas another disease attacks yellow-fleshed peaches far more than those with other coloured flesh. If, with all the aids of art, these slight differences make a great difference in cultivating the several varieties, assuredly, in a state of nature, where the trees would have to struggle with other trees and with a host of enemies, such differences would effectually settle which variety, whether a smooth or downy, a yellow or a purple fleshed fruit, should succeed.

In looking at many small points of difference between species, which, as far as our ignorance permits us to judge, seem to be quite unimportant, we must not forget that climate, food, &c., probably produce some slight and direct effect. It is, however, far more necessary to bear in mind that there are many unknown laws of correlation of growth, which when one part of the organisation is modified through variation, and the variations are accumulated by natural selection for the good of the being, will cause other modifications, often of the most unexpected nature....

Natural selection may modify and adapt the larva of an insect to a score of contingencies, wholly different from those which concern the mature insect. These modifications will no doubt affect, through the laws of correlation, the structure of the adult. ...So, conversely, modifications in the adult will probably often affect the structure of the larva; but in all cases natural selection will ensure that modifications...shall not be in the least degree injurious: for if they became so, they would cause the extinction of the species....

In order to make clear how, as I believe, natural selection acts, I must beg permission to give one or two imaginary illustrations. Let us take the case of a wolf, which preys on various animals, securing some by craft, some by strength, and some by fleetness; and let us suppose that the fleetest prey, a deer for instance, had from any change in the country increased in numbers, or that other prey had decreased in numbers, during that season of the year when the wolf was hardest pressed for food. I can under such circumstances see no reason to doubt that the swiftest and slimmest wolves would have the best chance of surviving and so be preserved or selected,—provided always that they retained strength to master their prey at this or some other period of the year, when they were compelled to prey on other animals. I can see no more reason to doubt this, than that man can improve the fleetness of his greyhounds by careful and methodical selection, or by that unconscious selection which results from each man trying to keep the best dogs without any thought of modifying the breed. I may add, that, according to Mr. Pierce, there are two varieties of the wolf inhabiting the Catskill Mountains, in the United States, one with a light greyhound-like form, which pursues deer, and the other more bulky, with shorter legs, which more frequently attacks the shepherd's flocks....

Let us now take a more complex case. Certain plants excrete a sweet juice, apparently for the sake of eliminating something injurious from the sap: this is effected, by glands at the base of the stipules in some Leguminosæ, and at the back of the leaf of the common laurel. This juice, though small in quantity, is greedily sought by insects. Let us now suppose a little sweet juice or nectar to be excreted by the inner bases of the petals of a flower. In this case insects in seeking the nectar would get dusted with pollen, and would certainly often transport the pollen from one flower to the stigma of another flower. The flowers of two distinct individuals of the same species would thus get crossed; and the act of crossing, we have good reason to believe, would produce very vigorous seedlings, which consequently would have the best chance of flourishing and surviving. Some of these seedlings would probably inherit the nectar-excreting power. Those individual flowers which had the largest glands or nectaries, and which excreted most nectar, would be oftenest visited by insects, and would be oftenest crossed; and so in the long-run would gain the upper hand. These flowers, also, which had their stamens and pistils placed, in relation to the size and habits of the particular insects which visited them, so as to favour in any degree the transportal of their pollen from flower to flower, would likewise be favoured or selected. We might have taken the case of insects visiting flowers for the sake of collecting pollen instead of nectar; and as pollen is formed for the sole object of fertilisation, its destruction appears a simple loss to the plant; yet if a little pollen were carried, at first occasionally and then habitually, by the pollen-devouring insects from flower to flower, and a cross thus effected, although nine-tenths of the pollen were destroyed it might still be a great gain to the plant; and those individuals which produced more and more pollen, and had larger and larger anthers, would be selected.

Let us now turn to the nectar-feeding insects in our imaginary case; we may suppose the plant of which we have been slowly increasing the nectar by continued selection, to be a common plant; and that certain insects depended in main part on its nectar for food. I could give many facts, showing how anxious bees are to save time: for instance, their habit of cutting holes and sucking the nectar at the bases of certain flowers, which they can, with a very little more trouble, enter by the mouth. Bearing such facts in mind, I can see no reason to doubt that an accidental deviation in the size and form of the body, or in the curvature and length of the proboscis, &c., far too slight to be appreciated by us, might profit a bee or other insect, so that an individual so characterised would be able to obtain its food more quickly, and so have a better chance of living and bearing descendants.... Thus I can understand how a flower and a bee might slowly become, either simultaneously or one after the other, modified and adapted in the most perfect manner to each other, by the continued preservation of individuals presenting mutual and slightly favourable deviations of structure.

I am well aware that this doctrine of natural selection, exemplified in the above imaginary instances, is open to the same objections which were first urged against Sir Charles Lyell's noble views on "the modern changes of the earth, as illustrative of geology"; but we now seldom hear the action, for instance, of the coast-waves, called a trifling and insignificant cause, when applied to the excavation of gigantic valleys or to the formation of the longest lines of inland cliffs. Natural selection can act only by the preservation and accumulation of infinitesimally small inherited modifications, each profitable to the preserved being; and as modern geology has almost banished such views as the excavation of a great valley by a single diluvial wave, so will natural selection, if it be a true principle, banish the belief of the continued creation of new organic beings, or of any great and sudden modification in their structure.

THE DESCENT OF MAN

By Charles Darwin, M.A., F.R.S.

(London: 1871)

Chapter I. The Evidence of the Descent of Man from some Lower Form.

...*The Bodily Structure of Man.* It is notorious that man is constructed on the same general type or model with other mammals. All the bones in his skeleton can be compared with corresponding bones in a monkey, bat, or seal. So it is with his muscles, nerves, blood-vessels and internal viscera. The brain, the most important of all the organs, follows the same law, as shewn by Huxley and other anatomists. Bischoff, who is a hostile witness, admits that every chief fissure and fold in the brain of man has its analogy in that of the orang; but he adds that at no period of development do their brains perfectly agree;

nor could this be expected, for otherwise their mental powers would have been the same....But it would be superfluous here to give further details on the correspondence between man and the higher mammals in the structure of the brain and all other parts of the body.

It may, however, be worth while to specify a few points, not directly or obviously connected with structure, by which this correspondence or relationship is well shewn.

Man is liable to receive from the lower animals, and to communicate to them, certain diseases, as hydrophobia, variola, the glanders, etc., and this fact proves the close similarity of their tissues and blood, both in minute structure and composition, far more plainly than does their comparison under the best microscope, or by the aid of the best chemical analysis....

Man is infested with internal parasites, sometimes causing fatal effects; and is plagued by external parasites, all of which belong to the same genera or families with those infesting other mammals....

The whole process of that most important function, the reproduction of the species, is strikingly the same in all mammals, from the first act of courtship by the male, to the birth and nurturing of the young. Monkeys are born in almost as helpless a condition as our own infants: and in certain genera the young differ fully as much in appearance from the adults, as do our children from their full-grown parents. It has been urged by some writers, as an important distinction, that with man the young arrive at maturity at a much later age than with any other animal: but if we look to the races of mankind which inhabit tropical countries the difference is not great, for the orang is believed not to be adult till the age of from ten to fifteen years. Man differs from woman in size, bodily strength, hairyness, &c., as well as in mind, in the same manner as do the two sexes of many mammals. It is, in short, scarcely possible to exaggerate the close correspondence in general structure, in the minute structure of the tissues, in chemical composition and in constitution, between man and the higher animals, especially the anthropomorphous apes.

Embryonic Development. Man is developed from an ovule, about the 125th of an inch in diameter, which differs in no respect from the ovules of other animals. The embryo itself at a very early period can hardly be distinguished from that of other members of the vertebrate kingdom. At this period the arteries run in arch-like branches, as if to carry the blood to branchiæ which are not present in the higher vertebrata, though the slits on the sides of the neck still remain,...marking their former position. At a somewhat later period, when the extremities are developed, "the feet of lizards and mammals," as the illustrious Von Baer remarks, "the wings and feet of birds, no less than the hands and feet of man, all arise from the same fundamental form." "It is," says Prof. Huxley, "quite in the later stages of development that the young human being presents marked differences from the young ape, while the latter departs as much from the dog in its developments, as the man does. Startling as this last assertion may appear to be, it is demonstrably true."...

After the foregoing statements made by such high authorities, it would be superfluous on my part to give a number of borrowed details, shewing that the embryo of man closely resembles that of other mammals. It may, however, be added, that the human embryo likewise resembles in various points of structure. certain low forms when adult. For instance, the heart at first exists as a simple pulsating vessel; the excreta are voided through a cloacal passage; and the os coccyx projects like a true tail, "extending considerably beyond the rudimentary legs." In the embryos of all air-breathing vertebrates, certain glands, called the corpora Wolffiana, correspond with, and act like the kidneys of mature fishes. Even at a later embryonic period, some striking resemblances between man and the lower animals may be observed. Bischoff says that the convolutions of the brain in a human fœtus at the end of the seventh month reach about the same stage of development as in a baboon when adult. The great toe, as Prof. Owen remarks, "which forms the fulcrum when standing or walking, is perhaps the most characteristic peculiarity in the human structure," but in an embryo, about an inch in length, Prof. Wyman found "that the great toe was shorter than the others; and, instead of being parallel to them, projected at an angle from the side of the foot, thus corresponding with the permanent condition of this part in the quadrumana." I will conclude with a quotation from Huxley, who after asking, does man originate in a different way from a dog, bird, frog or fish? says, "the reply is not doubtful for a moment;

without question, the mode of origin, and the early stages of development of man, are identical with those of the animals immediately below him in the scale: without a doubt in these respects, he is far nearer to apes than the apes are to the dog."

Rudiments....Not one of the higher animals can be named which does not bear some part in a rudimentary condition; and man forms no exception to the rule....Rudimentary organs are eminently variable; and this is partly intelligible, as they are useless, or nearly useless, and consequently are no longer subjected to natural selection. They often become wholly suppressed. When this occurs, they are nevertheless liable to occasional reappearance through reversion—a circumstance well worthy of attention.

...Rudiments of various muscles have been observed in many parts of the human body; and not a few muscles, which are regularly present in some of the lower animals can occasionally be detected in man in a greatly reduced condition. Every one must have noticed the power which many animals, especially horses, possess of moving or twitching their skin; and this is effected by the panniculus carnosus. Remnants of this muscle in an efficient state are found in various parts of our bodies; for instance, the muscle on the forehead, by which the eyebrows are raised....

Some few persons have the power of contracting the superficial muscles on their scalps; and these muscles are in a variable and partly rudimentary condition. M. A. de Candolle has communicated to me a curious instance of the long-continued persistence or inheritance of this power, as well as of its unusual development. He knows a family, in which one member, the present head of the family, could, when a youth, pitch several heavy books from his head by the movement of the scalp alone; and he won wagers by performing this feat. His father, uncle, grandfather, and his three children possess the same power to the same unusual degree. This family became divided eight generations ago into two branches; so that the head of the above-mentioned branch is cousin in the seventh degree to the head of the other branch. This distant cousin resides in another part of France; and on being asked whether he possessed the same faculty, immediately exhibited his power. This case offers a good illustration how persistently an absolutely useless faculty may be transmitted.

The sense of smell is of the highest importance to the greater number of mammals—to some, as the ruminants, in warning them of danger; to others, as the carnivora, in finding their prey; to others, again, as the wild boar, for both purposes combined. But the sense of smell is of extremely slight service, if any, even to savages, in whom it is much more highly developed than in the civilised races. It does not warn them of danger, nor guide them to their food; nor does it prevent the Esquimaux from sleeping in the most fetid atmosphere, nor many savages from eating half-putrid meat. Those who believe in the principle of gradual evolution, will not readily admit that this sense in its present state was originally acquired by man, as he now exists. No doubt he inherits the power in an enfeebled and so far rudimentary condition, from some early progenitor, to whom it was highly serviceable, and by whom it was continually used. We can thus perhaps understand how it is, as Dr Maudsley has truly remarked, that the sense of smell in man "is singularly effective in recalling vividly the ideas and images of forgotten scenes and places"; for we see in those animals, which have this sense highly developed, such as dogs and horses, that old recollections of persons and places are strongly associated with their odour.

Man differs conspicuously from all the other Primates in being almost naked. But a few short straggling hairs are found over the greater part of the body in the male sex, and fine down on that of the female sex. There can be little doubt that the hairs thus scattered over the body are the rudiments of the uniform hairy coat of the lower animals....

It appears as if the posterior molar or wisdom-teeth were tending to become rudimentary in the more civilised races of man. These teeth are rather smaller than the other molars, as is likewise the case with the corresponding teeth in the chimpanzee and orang; and they have only two separate fangs. They do not cut through the gums till about the seventeenth year, and I have been assured by dentists that they are much more liable to decay, and are earlier lost, than the other teeth. It is also remarkable that they are much more liable to vary both in structure and in the period of their development, than the other teeth. In the Melanian races, on the other hand, the wisdom-teeth are usually furnished with three separate fangs, and are generally sound; they also differ from the other molars in size less than in the Caucasian races. Prof. Schaffhausen accounts for this difference between the races by "the posterior dental portion of the jaw being always shortened" in those that are civilised, and this shortening may, I presume, be safely attributed to civilised men habitually feeding on soft, cooked food, and thus using their jaws less....

With respect to the alimentary canal, I have met with an account of only a single rudiment, namely the vermiform appendage of the cæcum. The cæcum is a branch or diverticulum of the intestine, ending in a cul-de-sac, and is extremely long in many of the lower vegetable-feeding mammals. In the marsupial koala it is actually more than thrice as long as the whole body. It is sometimes produced into a long gradually-tapering point and is sometimes constricted in parts. It appears as if, in consequence of changed diet or habits, the cæcum had become much shortened in various animals, the vermiform appendage being left as a rudiment of the shortened part. That this appendage is a rudiment, we may infer from its small size, and from the evidence which Prof. Canestrini has collected of its variability in man. It is occasionally quite absent, or again is largely developed. The passage is sometimes completely closed for half or two-thirds of its length, with the terminal part consisting of a flattened solid expansion. In the orang this appendage is long and convoluted; in man it arises from the end of the short cæcum, and is commonly from four to five inches in length, being only about the third of an inch in diameter. Not only is it useless, but it is sometimes the cause of death, of which fact I have lately heard two instances; this is due to small hard bodies, such as seeds, entering the passage, and causing inflammation....

The os coccyx in man, though functionless as a tail, plainly represents this part in other vertebrate animals. At an early embryonic period it is free, and...projects beyond the lower extremities. In certain rare and anomalous cases, it has been known...to form a small external rudiment of a tail.

The bearing of the three great classes of facts now given is unmistakeable. But it would be superfluous here fully to recapitulate the line of argument given in detail in my "Origin of Species." The homological construction of the whole frame in the members of the same class is intelligible, if we admit their descent from a common progenitor, together with their subsequent adaptation to diversified conditions. On any other view, the similarity of pattern between the hand of a man or monkey, the foot of a horse, the flipper of a seal, the wing of a bat, &c., is utterly inexplicable. It is no scientific explanation to assert that they have all been formed on the same ideal plan. With respect to development, we can clearly understand, on the principle of variation supervening at a rather late embryonic period, and being inherited at a corresponding period, how it is that the embryos of wonderfully different forms should still retain, more or less perfectly, the structure of their common progenitor. No other explanation has ever been given of the marvellous fact that the embryos of a man, dog, seal, bat, reptile, &c., can at first hardly be distinguished from each other. In order to understand the existence of rudimentary organs, we have only to suppose that a former progenitor possessed the parts in question in a perfect state, and that under changed habits of life they became greatly reduced, either from simple disuse, or through the natural selection of those individuals which were least encumbered with a superfluous part....

Thus we can understand how it has come to pass that man and all other vertebrate animals have been constructed on the same general model, why they pass through the same early stages of development, and why they retain certain rudiments in common. Consequently we ought frankly to admit their community of descent; to take any other view, is to admit that our own structure, and that of all the animals around us, is a mere snare laid to entrap our judgment. This conclusion is greatly strengthened, if we look to the members of the whole animal series and consider the evidence derived from their affinities or classification, their geographical distribution and geological succession. It is only our natural prejudice, and that arrogance which made our forefathers declare that they were descended from demi-gods, which leads us to demur to this conclusion. But the time will before long come, when it will be thought wonderful, that naturalists, who were well acquainted with the comparative structure and development of man, and other mammals, should have believed that each was the work of a separate act of creation.

THE LAWS OF HEREDITY: MENDEL

EXPERIMENTS on heredity had been carried out before Darwin's time, but to little effect, especially as many hybrids were found to be infertile. Darwin writes in the *Origin of Species* "the laws governing inheritance are for the most part unknown." The simplest of these laws were discovered about 1860 by Mendel, whose work was, however, completely neglected till 1900.

Gregor Johann Mendel was born in Austrian Silesia in 1822. The son of a small peasant proprietor, he received a good education, partly through the sacrifice of her dowry by one of his sisters. At the age of 21, he entered the Augustinian monastery at Brünn. After being ordained priest, he studied natural science at Vienna for three years, and then returned to Brünn to teach. He had wide interests, carrying out experiments on plants and on bees in the cloister garden, observing sunspots, and writing on meteorology. The results of his hybridizations were published in the Proceedings of the local scientific society, and the general neglect of his work much disappointed him. He was elected abbot of the monastery in 1868, when his researches practically ended, and after some busy years, partly spent in political and racial controversy, he died in 1884.

His successful elucidation of some of the laws of heredity was due to his choice of a suitable plant, and his careful tabulation of the characters of all the descendants of his hybrids and their statistical relations.

EXPERIMENTS IN PLANT-HYBRIDISATION

By GREGOR MENDEL

(From the Proceedings of the Brünn Natural History Society: read at the Meetings of the 8th February and 8th March, 1865.)

(Translation taken from *Mendel's Principles of Heredity*, by W. Bateson, 1909.)

INTRODUCTORY REMARKS.

EXPERIENCE of artificial fertilisation, such as is effected with ornamental plants in order to obtain new variations in colour, has led to the experiments which will here be discussed. The striking regularity with which the same hybrid forms always reappeared whenever fertilisation took place between the same species induced further experiments to be undertaken, the object of which was to follow up the developments of the hybrids in their progeny....

That, so far, no generally applicable law governing the formation and development of hybrids has been successfully formulated can hardly be wondered at by anyone who is acquainted with the extent of the task, and can appreciate the difficulties with which experiments of this class have to contend. A final decision can only be arrived at when we shall have before us the results of detailed experiments made on plants belonging to the most diverse orders.

Those who survey the work done in this department will arrive at the conviction that, among all the numerous experiments made, not one has been carried out to such an extent and in such a way as to make it possible to determine the number of different forms under which the offspring of hybrids appear, or to arrange these forms with certainty according to their separate generations, or definitely to ascertain their statistical relations. It requires indeed some courage to undertake a labour of such far-reaching extent; this appears, however, to be the only right way by which we can finally reach the solution of a question the importance of which cannot be over-estimated in connection with the history of the evolution of organic forms.

The paper now presented records the results of such a detailed experiment. This experiment was practically confined to a small plant group, and is now, after eight years' pursuit, concluded in all essentials. Whether the plan upon which the separate experiments were conducted and carried out was the best suited to attain the desired end is left to the friendly decision of the reader.

Selection of the Experimental Plants.

...The selection of the plant group which shall serve for experiments of this kind must be made with all possible care if it be desired to avoid from the outset every risk of questionable results.

The experimental plants must necessarily—

1. Possess constant differentiating characters.

2. The hybrids of such plants must, during the flowering period, be protected from the influence of all foreign pollen, or be easily capable of such protection.

The hybrids and their offspring should suffer no marked disturbance in their fertility in the successive generations.

Accidental impregnation by foreign pollen, if it occurred during the experiments and were not recognized, would lead to entirely erroneous conclusions. Reduced fertility or entire sterility of certain forms, such as occurs in the offspring of many hybrids, would render the experiments very difficult or entirely frustrate them. In order to discover the relations in which the hybrid forms stand towards each other and also towards their progenitors, it appears to be necessary that all members of the series developed in each successive generation should be, *without exception*, subjected to observation.

At the very outset special attention was devoted to the *Leguminosæ* on account of their peculiar floral structure. Experiments which were made with several members of this family led to the result that the genus *Pisum* was found to possess the necessary qualifications.

Some thoroughly distinct forms of this genus possess characters which are constant, and easily and certainly recognizable, and when their hybrids are mutually crossed they yield perfectly fertile progeny. Furthermore, a disturbance through foreign pollen cannot easily occur, since the fertilising organs are closely packed inside the keel and the anther bursts within the bud, so that the stigma becomes covered with pollen even before the flower opens. This circumstance is of especial importance. As additional advantages worth mentioning, there may be cited the easy culture of these plants in the open ground and in pots, and also their relatively short period of growth. Artificial fertilisation is certainly a somewhat elaborate process, but nearly always succeeds. For this purpose the bud is opened before it is perfectly developed, the keel is removed, and each stamen carefully extracted by means of forceps, after which the stigma can at once be dusted over with the foreign pollen....

Division and Arrangement of the Experiments.

If two plants which differ constantly in one or several characters be crossed, numerous experiments have demonstrated that the common characters are transmitted unchanged to the hybrids and their progeny; but each pair of differentiating characters, on the other hand, unite in the hybrid to form a new character, which in the progeny of the hybrid is usually variable. The object of the experiment was to observe these variations in the case of each pair of differentiating characters, and to deduce the law according to which they appear in the successive generations....

The characters which were selected for experiment relate:

1. To the *difference in the form of the ripe seeds.* These are either round or roundish; the depressions, if any, occur on the surface, being always only shallow; or they are irregularly angular and deeply wrinkled (*P. quadratum*).

2. To the *difference in the colour of the seed albumen* (endosperm). The albumen of the ripe seeds is either pale yellow, bright yellow and orange coloured, or it possesses a more or less intense green tint. This difference of colour is easily seen in the seeds as their coats are transparent.

3. To the *difference in the colour of the seed-coat.* This is either white, with which character white flowers are constantly correlated, or it is grey, grey-brown, leather-brown, with or without violet spotting, in which case the colour of the standards is violet, that of the wings purple, and the stem in the axils of the leaves is of a reddish tint. The grey seed-coats become dark brown in boiling water.

4. To the *difference in the form of the ripe pods.* These are either simply inflated, not contracted in places; or they are deeply constricted beween the seeds and more or less wrinkled (*P. saccharatum*).

5. To the *difference in the colour of the unripe pods.* They are either light to dark green, or vividly yellow, in which colouring the stalks, leaf-veins, and calyx participate.

6. To the *difference in the position of the flowers.* They are either axial, that is, distributed along the main stem; or they are terminal, that is, bunched at the top of the stem and arranged almost in a false umbel; in this case the upper part of the stem is more or less widened in section (*P. umbellatum*).

7. To the *difference in the length of the stem.* The length of the stem is very various in some forms; it is, however, a constant character for each, in so far that healthy plants, grown in the same soil, are only subject to unimportant variations in this character.

In experiments with this character, in order to be able to

discriminate with certainty, the long axis of 6 to 7 ft. was always crossed with the short one of $\frac{3}{4}$ ft. to $1\frac{1}{2}$ ft.

Each two of the differentiating characters enumerated above were united by cross-fertilisation. There were made for the

1st	trial	60	fertilisations	on	15	plants.
2nd	"	58	"	"	10	"
3rd	"	35	"	"	10	"
4th	"	40	"	"	10	"
5th	"	23	"	"	5	"
6th	"	34	"	"	10	"
7th	"	37	"	"	10	"

From a larger number of plants of the same variety only the most vigorous were chosen for fertilisation. Weakly plants always afford uncertain results, because even in the first generation of hybrids, and still more so in the subsequent ones, many of the offspring either entirely fail to flower or only form a few and inferior seeds.

Furthermore, in all the experiments reciprocal crossings were effected in such a way that each of the two varieties which in one set of fertilisations served as seed-bearer in the other set was used as the pollen plant....

The Forms of the Hybrids.

Experiments which in previous years were made with ornamental plants have already afforded evidence that the hybrids, as a rule, are not exactly intermediate between the parental species....This is...the case with the Pea-hybrids. In the case of each of the seven crosses the hybrid-character resembles that of one of the parental forms so closely that the other either escapes observation completely or cannot be detected with certainty. This circumstance is of great importance in the determination and classification of the forms under which the offspring of the hybrids appear. Henceforth in this paper those characters which are transmitted entire, or almost unchanged in the hybridisation, and therefore in themselves constitute the characters of the hybrid, are termed the *dominant*, and those which become latent in the process *recessive*. The expression "recessive" has been chosen because the characters thereby designated withdraw or entirely disappear in the hybrids, but nevertheless reappear unchanged in their progeny, as will be demonstrated later on.

It was furthermore shown by the whole of the experiments that it is perfectly immaterial whether the dominant character belong to the seed-bearer or to the pollen-parent; the form of the hybrid remains identical in both cases....

Of the differentiating characters which were used in the experiments the following are dominant:

1. The round or roundish form of the seed with or without shallow depressions.

2. The yellow colour of the seed albumen.

3. The grey, grey-brown, or leather-brown colour of the seed-coat, in association with violet-red blossoms and reddish spots in the leaf axils.

4. The simply inflated form of the pod.

5. The green colouring of the unripe pod in association with the same colour in the stems, the leaf-veins and the calyx.

6. The distribution of the flowers along the stem.

7. The greater length of stem....

The First Generation [bred] from the Hybrids.

In this generation there reappear, together with the dominant characters, also the recessive ones with their peculiarities fully developed, and this occurs in the definitely expressed average proportion of three to one, so that among four plants of this generation three display the dominant character and one the recessive. This relates without exception to all the characters which were investigated in the experiments....*Transitional forms were not observed in any experiment....*

...The relative numbers which were obtained for each pair of differentiating characters are as follows:

Expt. 1. Form of seed.—From 253 hybrids 7,324 seeds were obtained in the second trial year. Among them were 5,474 round or roundish ones and 1,850 angular wrinkled ones. Therefrom the ratio 2·96 to 1 is deduced.

Expt. 2. Colour of albumen.—258 plants yielded 8,023 seeds, 6,022 yellow, and 2,001 green; their ratio, therefore, is as 3·01 to 1....

Expt. 3. Colour of the seed-coats.—Among 929 plants 705 bore violet-red flowers and grey-brown seed-coats, giving the proportion 3·15 to 1.

Expt. 4. Form of pods.—Of 1,181 plants 882 had them simply inflated, and in 299 they were constricted. Resulting ratio, 2·95 to 1.

Expt. 5. Colour of the unripe pods.—The number of trial plants was 580, of which 428 had green pods and 152 yellow ones. Consequently these stand in the ratio 2·82 to 1.

Expt. 6. Position of flowers.—Among 858 cases 651 had inflorescences axial and 207 terminal. Ratio, 3·14 to 1.

Expt. 7. Length of stem.—Out of 1,064 plants, in 787 cases the stem was long, and in 277 short. Hence a mutual ratio of 2·84 to 1. In this experiment the dwarfed plants were carefully lifted and transferred to a special bed. This precaution was necessary, as otherwise they would have perished through being overgrown by their tall relatives. Even in their quite young state they can be easily picked out by their compact growth and thick dark-green foliage.

If now the results of the whole of the experiments be brought together, there is found, as between the number of forms with the dominant and recessive characters, an average ratio of 2·98 to 1, or 3 to 1.

The dominant character can here have a *double signification*—viz. that of a parental character, or a hybrid character. In which of the two significations it appears in each separate case can only be determined by the following generation. As a parental character it must pass over unchanged to the whole of the offspring; as a hybrid-character, on the other hand, it must maintain the same behaviour as in the first generation.

The Second Generation [bred] from the Hybrids.

Those forms which in the first generation exhibit the recessive character do not further vary in the second generation as regards this character; they remain constant in their offspring.

It is otherwise with those which possess the dominant character in the first generation (bred from the hybrids). Of these *two*-thirds yield offspring which display the dominant and recessive characters in the proportion of 3 to 1, and thereby show exactly the same ratio as the hybrid forms, while only *one*-third remains with the dominant character constant.

The separate experiments yielded the following results:

Expt. 1. Among 565 plants which were raised from round seeds of the first generation, 193 yielded round seeds only, and remained therefore constant in this character; 372, however, gave both round and wrinkled seeds, in the proportion of 3 to 1. The number of the hybrids, therefore, as compared with the constants is 1·93 to 1.

Expt. 2. Of 519 plants which were raised from seeds whose albumen was of yellow colour in the first generation, 166 yielded exclusively yellow, whilst 353 yielded yellow and green seeds in the proportion of 3 to 1. There resulted, therefore, a division into hybrid and constant forms in the proportion of 2·13 to 1.

For each separate trial in the following experiments 100 plants were selected which displayed the dominant character in the first generation, and in order to ascertain the significance of this, ten seeds of each were cultivated.

Expt. 3. The offspring of 36 plants yielded exclusively grey-brown seed-coats, while of the offspring of 64 plants some had grey-brown and some had white.

Expt. 4. The offspring of 29 plants had only simply inflated pods; of the offspring of 71, on the other hand, some had inflated and some had constricted.

Expt. 5. The offspring of 49 plants had only green pods; of the offspring of 60 plants some had green, some yellow ones.

Expt. 6. The offspring of 33 plants had only axial flowers; of the offspring of 67, on the other hand, some had axial and some terminal flowers.

Expt. 7. The offspring of 28 plants inherited the long axis, and those of 72 plants some the long and some the short axis.

In each of these experiments a certain number of the plants came constant with the dominant character. For the determination of the proportion in which the separation of the forms with the constantly persistent character results, the two first experiments are of especial importance, since in these a larger number of plants can be compared. The ratios 1·93 to 1 and 2·13 to 1 gave together almost exactly the average ratio of 2 to 1. The sixth experiment gave a quite concordant result; in the others the ratio varies more or less, as was only to be expected in view of the smaller number of 100 trial plants. Experiment 5, which shows the greatest departure, was repeated, and then, in lieu of the ratio of 60 and 40, that of 65 and 35 resulted. *The average ratio of 2 to 1 appears, therefore, as fixed with certainty.* It is therefore demonstrated that, of those forms which possess the dominant character in the first generation, two-thirds have the hybrid-character, while one-third remains constant with the dominant character.

The ratio of 3 to 1, in accordance with which the distribution of the dominant and recessive characters results in the first generation, resolves itself therefore in all experiments into the ratio of 2 : 1 : 1 if the dominant character be differentiated according to its significance as a hybrid-character or as a parental one. Since the members of the first generation spring directly from the seed of the hybrids, *it is now clear that the hybrids form seeds having one or the other of the two differentiating characters, and of these one-half develop again the hybrid form, while the other half yield plants which remain constant and receive the dominant or the recessive characters in equal numbers.*

THE SUBSEQUENT GENERATIONS [BRED] FROM THE HYBRIDS.

The proportions in which the descendants of the hybrids develop and split up in the first and second generations presumably hold good for all subsequent progeny. Experiments 1 and 2 have already been carried through six generations, 3 and 7 through five, and 4, 5, and 6 through four, these experiments being continued from the third generations with a small number of plants, and no departure from the rule has been perceptible. The offspring of the hybrids separated in each generation in the ratio of 2 : 1 : 1 into hybrids and constant forms.

If *A* be taken as denoting one of the two constant characters, for instance the dominant, a, the recessive, and Aa the hybrid form in which both are conjoined, the expression

$$A + 2Aa + a$$

shows the terms in the series for the progeny of the hybrids of two differentiating characters....

The Offspring of Hybrids in which several Differentiating Characters are associated.

In the experiments above described plants were used which differed only in one essential character. The next task consisted in ascertaining whether the law of development discovered in these applied to each pair of differentiating characters when several diverse characters are united in the hybrid by crossing. As regards the form of the hybrids in these cases, the experiments showed throughout that this invariably more nearly approaches to that one of the two parental plants which possesses the greater number of dominant characters....Should one of the two parental types possess only dominant characters, then the hybrid is scarcely or not at all distinguishable from it.

Two experiments were made with a considerable number of plants. In the first experiment the parental plants differed in the form of the seed and in the colour of the albumen; in the second in the form of the seed, in the colour of the albumen, and in the colour of the seed-coats. Experiments with seed-characters give the result in the simplest and most certain way.

In addition, further experiments were made with a smaller number of experimental plants in which the remaining characters by twos and threes were united as hybrids; all yielded approximately the same results. There is therefore no doubt that for the whole of the characters involved in the experiments the principle applies that *the offspring of the hybrids in which several essentially different characters are combined exhibit the terms of a series of combinations, in which the developmental series for each pair of differentiating characters are united.* It is demonstrated at the same time that *the relation of each pair of different characters in hybrid union is independent of the other differences in the two original parental stocks....*

All constant combinations which in Peas are possible by the

combination of the said seven differentiating characters were actually obtained by repeated crossing....Thereby is...given the practical proof *that the constant characters which appear in the several varieties of a group of plants may be obtained in all the associations which are possible according to the [mathematical] laws of combination, by means of repeated artificial fertilisation....*

If we endeavour to collate in a brief form the results arrived at, we find that those differentiating characters, which admit of easy and certain recognition in the experimental plants, all behave exactly alike in their hybrid associations. The offspring of the hybrids of each pair of differentiating characters are, one-half, hybrid again, while the other half are constant in equal proportions having the characters of the seed and pollen parents respectively. If several differentiating characters are combined by cross-fertilisation in a hybrid, the resulting offspring form the terms of a combination series in which the combination series for each pair of differentiating characters are united.

The uniformity of behaviour shown by the whole of the characters submitted to experiment permits, and fully justifies, the acceptance of the principle that a similar relation exists in the other characters which appear less sharply defined in plants, and therefore could not be included in the separate experiments....

The Reproductive Cells of the Hybrids.

The results of the previously described experiments led to further experiments, the results of which appear fitted to afford some conclusions as regards the composition of the egg and pollen cells of hybrids. An important clue is afforded in *Pisum* by the circumstance that among the progeny of the hybrids constant forms appear, and that this occurs, too, in respect of all combinations of the associated characters. So far as experience goes, we find it in every case confirmed that constant progeny can only be formed when the egg cells and the fertilising pollen are of like character, so that both are provided with the material for creating quite similar individuals, as is the case with the normal fertilisation of pure species. We must therefore regard it as certain that exactly similar factors must be at work also in the production of the constant forms in the hybrid plants. Since the various constant forms are produced in *one* plant, or even in one flower of a plant, the conclusion appears logical that in the ovaries of the hybrids there are formed as many sorts of egg cells, and in the anthers as many sorts of pollen cells, as there are possible constant combination forms, and that these egg and pollen cells agree in their internal composition with those of the separate forms.

In point of fact it is possible to demonstrate theoretically that this hypothesis would fully suffice to account for the development of the hybrids in the separate generations, if we might at the same time assume that the various kinds of egg and pollen cells were formed in the hybrids on the average in equal numbers.

In order to bring these assumptions to an experimental proof, the following experiments were designed. Two forms which were constantly different in the form of the seed and the colour of the albumen were united by fertilisation.

If the differentiating characters are again indicated as *A*, *B*, *a*, *b*, we have:

AB, seed parent;	*ab*, pollen parent;
A, form round;	*a*, form wrinkled;
B, albumen yellow.	*b*, albumen green.

The artificially fertilised seeds were sown together with several seeds of both original stocks, and the most vigorous examples were chosen for the reciprocal crossing. There were fertilised:

1. The hybrids with the pollen of *AB*.
2. The hybrids " " " " *ab*.
3. *AB* " " " " the hybrids.
4. *ab* " " " " the hybrids.

For each of these four experiments the whole of the flowers on three plants were fertilised. If the above theory be correct, there must be developed on the hybrids egg and pollen cells of the forms *AB*, *Ab*, *aB*, *ab*, and there would be combined:

1. The egg cells *AB*, *Ab*, *aB*, *ab* with the pollen cells *AB*.
2. The egg cells *AB*, *Ab*, *aB*, *ab* with the pollen cells *ab*.
3. The egg cells *AB* with the pollen cells *AB*, *Ab*, *aB*, *ab*.
4. The egg cells *ab* with the pollen cells *AB*, *Ab*, *aB*, *ab*.

From each of these experiments there could then result only the following forms:

1. *AB*, *ABb*, *AaB*, *AaBb*.
2. *AaBb*, *Aab*, *aBb*, *ab*.
3. *AB*, *ABb*, *AaB*, *AaBb*.
4. *AaBb*, *Aab*, *aBb*, *ab*.

If, furthermore, the several forms of the egg and pollen cells of the hybrids were produced on an average in equal numbers, then in each experiment the said four combinations should stand in the same ratio to each other. A perfect agreement in the numerical relations was, however, not to be expected, since in each fertilisation, even in normal cases, some egg cells remain undeveloped or subsequently die, and many even of the well-formed seeds fail to germinate when sown....

The first and second experiments had primarily the object of proving the composition of the hybrid egg cells, while the third and fourth experiments were to decide that of the pollen cells. As is shown by the above demonstration the first and third experiments and the second and fourth should produce precisely the same combinations, and even in the second year the result should be partially visible in the form and colour of the artificially fertilised seed. In the first and third experiments the dominant characters of form and colour, *A* and *B*, appear in each union,...partly constant and partly in hybrid union with the recessive characters *a* and *b*, for which reason they must impress their peculiarity upon the whole of the seeds. All seeds should therefore appear round and yellow, if the theory be justified. In the second and fourth experiments, on the other hand, one union is hybrid in form and in colour, and consequently the seeds are round and yellow; another is hybrid in form, but constant in the recessive character of colour, whence the seeds are round and green; the third is constant in the recessive character of form but hybrid in colour, consequently the seeds are wrinkled and yellow; the fourth is constant in both recessive characters, so that the seeds are wrinkled and green. In both these experiments there were consequently four sorts of seed to be expected—viz. round and yellow, round and green, wrinkled and yellow, wrinkled and green.

The crop fulfilled these expectations perfectly. There were obtained in the

1st Experiment, 98 exclusively round yellow seeds;
3rd " 94 " " " "

In the 2nd Experiment, 31 round and yellow, 26 round and green, 27 wrinkled and yellow, 26 wrinkled and green seeds.

In the 4th Experiment, 24 round and yellow, 25 round and green, 22 wrinkled and yellow, 27 wrinkled and green seeds....

In a further experiment the characters of flower-colour and length of stem were experimented upon....For the characters of form of pod, colour of pod, and position of flowers experiments were also made on a small scale, and results obtained in perfect agreement. All combinations which were possible through the union of the differentiating characters duly appeared, and in nearly equal numbers.

Experimentally, therefore, the theory is confirmed that *the pea hybrids form pollen and egg cells which, in their constitution, represent in equal numbers all constant forms which result from the combination of the characters united in fertilisation.*

The difference of the forms among the progeny of the hybrids, as well as the respective ratios of the numbers in which they are observed, find a sufficient explanation in the principle above deduced. The simplest case is afforded by the developmental series of each pair of differentiating characters. This series is represented by the expression $A + 2Aa + a$, in which A and a signify the forms with constant differentiating characters, and Aa the hybrid form of both. It includes in three different classes four individuals. In the formation of these, pollen and egg cells of the form A and a take part on the average equally in the fertilisation, hence each form occurs twice, since four individuals are formed. There participate consequently in the fertilisation

The pollen cells $A + A + a + a$.
The egg cells $A + A + a + a$.

It remains, therefore, purely a matter of chance which of the two sorts of pollen will become united with each separate egg cell. According, however, to the law of probability, it will always happen, on the average of many cases, that each pollen form A and a will unite equally often with each egg cell form

A and *a*, consequently one of the two pollen cells *A* in the fertilisation will meet with the egg cell *A*, and the other with an egg cell *a*, and so likewise one pollen cell will unite with an egg cell *A*, and the other with egg cell *a*.

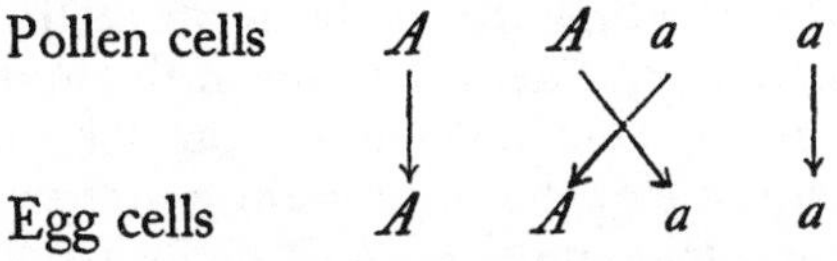

The results of the fertilisation may be made clear by putting the signs for the conjoined egg and pollen cells in the form of fractions, those for the pollen cells above and those for the egg cells below the line. We then have

$$\frac{A}{A} + \frac{A}{a} + \frac{a}{A} + \frac{a}{a}.$$

In the first and fourth term the egg and pollen cells are of like kind, consequently the product of their union must be constant, viz. *A* and *a*; in the second and third, on the other hand, there again results a union of the two differentiating characters of the stocks, consequently the forms resulting from these fertilisations are identical with those of the hybrid from which they sprang. *There occurs accordingly a repeated hybridisation.* This explains the striking fact that the hybrids are able to produce, besides the two parental forms, offspring which are like themselves; $\frac{A}{a}$ and $\frac{a}{A}$ both give the same union *Aa*, since, as already remarked above, it makes no difference in the result of fertilisation to which of the two characters the pollen or egg cells belong. We may write then

$$\frac{A}{A} + \frac{A}{a} + \frac{a}{A} + \frac{a}{a} = A + 2Aa + a.$$

This represents the average result of the self-fertilisation of the hybrids when two differentiating characters are united in them. In individual flowers and in individual plants, however, the ratios in which the forms of the series are produced may suffer not inconsiderable fluctuations. Apart from the fact that the numbers in which both sorts of egg cells occur in the seed vessels can only be regarded as equal on the average, it remains purely a matter of chance which of the two sorts of pollen may fertilise each separate egg cell. For this reason the separate values must necessarily be subject to fluctuations, and there are even extreme cases possible, as were described earlier in connection with the experiments on the form of the seed and the colour of the albumen. The true ratios of the numbers can only be ascertained by an average deduced from the sum of as many single values as possible; the greater the number, the more are merely chance effects eliminated.

The law of combination of different characters which governs the development of the hybrids finds therefore its foundation and explanation in the principle enunciated, that the hybrids produce egg cells and pollen cells which in equal numbers represent all constant forms which result from the combinations of all the characters brought together in fertilisation.

THE CHROMOSOME THEORY OF HEREDITY

During the nineteenth century, the minute structure of the living organism was studied, and all the various parts—brain, liver, muscle; leaf, flower and root, were found to be made up of small units or cells multiplying by division. Each cell contains a central spherical body or nucleus; within this is a network of fibres, which at each division of the cell resolves itself into a definite number of threads or chromosomes. The behaviour of the chromosomes led some observers to see in them a possible mechanism for Mendelian heredity. The theory received support and amplification from the work of Professor T. H. Morgan, of Columbia University, and his colleagues, on the heredity of a quick-breeding fly. Accepting the chomosomes as the bearers of Mendelian factors, these workers feel justified in locating factors in definite chromosomes and even in mapping the relative position of factors in one chromosome. Though not intrinsically more important than much other work now in progress, it is of interest to us as supplementing Mendel's own. There follows a brief introduction to the subject by Professor Morgan, from his book *The Mechanism of Mendelian Heredity*.

THE MECHANISM OF MENDELIAN HEREDITY

By T. H. Morgan and others

New York, 1915.

Chapter I. *Mendelian Segregation and the Chromosomes.*

Mendel's law was announced in 1865. Its fundamental principle is very simple. *The units contributed by two parents separate in the germ cells of the offspring without having had any influence on each other.* For example, in a cross between yellow-seeded and green-seeded peas, one parent contributes to the offspring a unit for yellow and the other parent contributes a unit for green. These units separate in the ripening of the germ cells of the offspring so that half of the germ cells are yellow bearing and half are green bearing. This separation occurs both in the eggs and in the sperm.

Mendel did not know of any mechanism by which such a process could take place. In fact, in 1865 very little was known about the ripening of the germ cells. But in 1900, when Mendel's long-forgotten discovery was brought to light once more, a mechanism had been discovered that fulfils exactly the Mendelian requirements of pairing and separation.

The sperm of every species of animal or plant carries a definite number of bodies called chromosomes. The egg carries the same number. Consequently, when the sperm unites with the egg, the fertilized egg will contain the double number of chromosomes. For each chromosome contributed by the sperm there is a corresponding chromosome contributed by the egg, i.e., there are two chromosomes of each kind, which together constitute a pair.

When the egg divides (Fig. 1, *a–d*), every chromosome splits into two chromosomes, and these daughter chromosomes then move apart, going to opposite poles of the dividing cell (Fig. 1, *c*). Thus each daughter cell (Fig. 1, *d*) receives one of the daughter chromosomes formed from each original chromosome. The same process occurs in all cell divisions, so that all the cells of the animal or plant come to contain the double set of chromosomes.

The germ cells also have at first the double set of chromosomes, but when they are ready to go through the last stages of their transformation into sperm or eggs the chromosomes unite in pairs (Fig. 1, *e*). Then follows a different kind of division

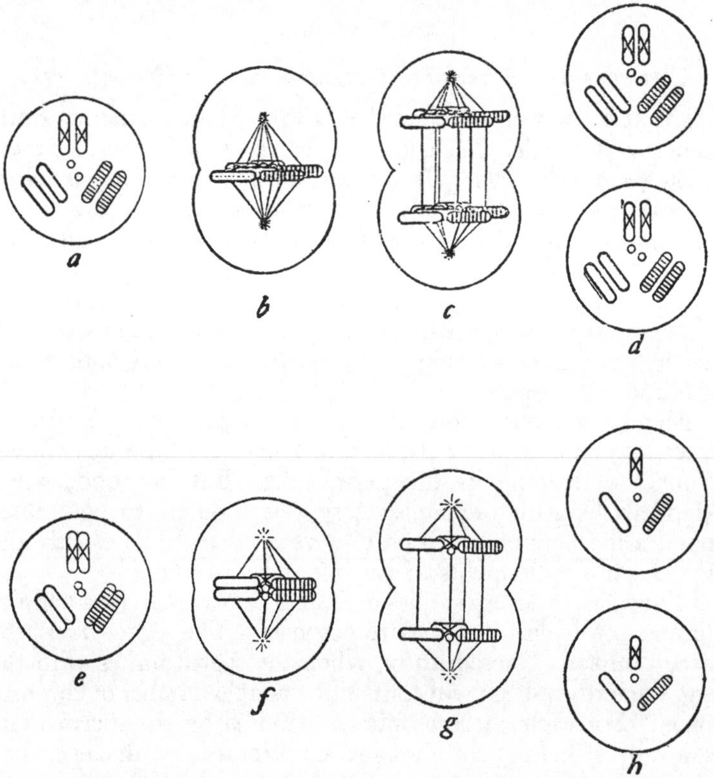

Fig. 1. In the upper line, four stages in the division of the egg (or of a body cell) are represented. Every chromosome divides when the cell divides. In the lower line the "reduction division" of a germ cell, after the chromosomes have united in pairs, is represented. The members of each of the four pairs of chromosomes separate from each other at this division.

(Fig. 1, *f*), at which the chromosomes do not split, but the members of each pair of chromosomes separate and each member goes into one of the daughter cells (Fig. 1, *g*, *h*). As a result each mature germ cell receives one or the other member of every pair of chromosomes and the number is reduced to half. Thus the behaviour of the chromosomes parallels the behaviour of the Mendelian units, for in the germ cells each unit derived from the father separates from the corresponding unit derived from the mother. These units will henceforth be spoken of as factors; the two factors of a pair are called allelomorphs of each other. Their separation in the germ cells is called segregation.

The possibility of explaining Mendelian phenomena by means of the manœuvres of the chromosomes seems to have occurred to more than one person, but Sutton was the first to present the idea in the form in which we recognize it to-day. Moreover, he not only called attention to the fact above mentioned, that both chromosomes and hereditary factors undergo segregation, but showed that the parallelism between their methods of distribution goes even further than this. Mendel had found that when the inheritance of more than one pair of factors is followed, the different pairs of factors segregate independently of one another. Thus in a cross of a pea having both green seeds and tall stature with a pea having yellow seeds and short stature, the fact that a germ cell receives a particular member of one pair (e.g. yellow) does not determine which member of the other pair it receives; it is as likely to receive the tall as the short. Sutton pointed out that in the same way the segregation of one pair of chromosomes is probably independent of the segregation of the other pairs.

It was obvious from the beginning, however, that there was one essential requirement of the chromosome view, namely, that all the factors carried by the *same* chromosome should tend to remain together. Therefore, since the number of inheritable characters may be large in comparison with the number of pairs of chromosomes, we should expect actually to find not only the independent behaviour of pairs, but also cases in which characters are linked together in groups in their inheritance. Even in species where a limited number of Mendelian units are known, we should still expect to find some of them in groups.

In 1906 Bateson and Punnett made the discovery of linkage, which they called gametic coupling. They found that when a sweet pea with factors for purple flowers and long pollen grains was crossed to a pea with factors for red flowers and round pollen grains, the two factors that came from the same parent tended to be inherited together. Here was the first case that gave the sort of result that was to be expected if factors were in chromosomes, although this relation was not pointed out at the time. In the same year, however, Lock called attention to the possible relation between the chromosome hypothesis and linkage.

In other groups a few cases of coupling became known, but nowhere had the evidence been sufficiently ample or sufficiently studied to show how frequently coupling occurs. Since 1910, however, in the fruit fly, Drosophila ampelophila, a large number of new characters have appeared by mutation, and so rapidly does the animal reproduce that in a relatively short time the inheritance of more than a hundred characters has been studied. It became evident very soon that these characters are inherited in groups. There is one great group of characters that are sex linked. There are two other groups of characters slightly greater in number. Finally a character appeared that did not belong to any of the other groups, and a year later still another character appeared that was linked to the last one but was independent of all the other groups. Hence there are four groups of characters in Drosophila....

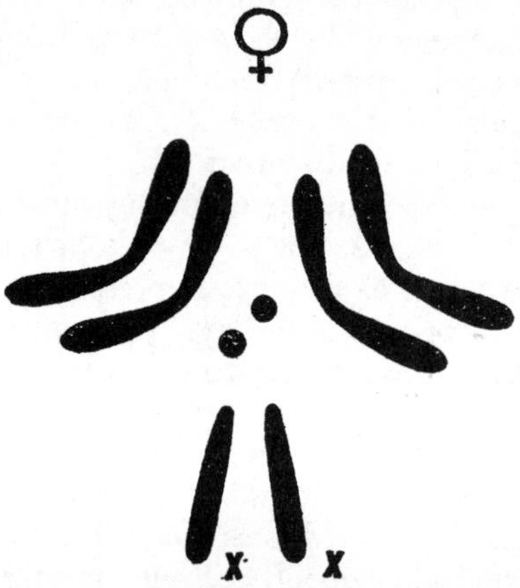

Fig. 2. Diagram of female group (duplex) of chromosomes of Drosophila ampelophila showing the four pairs of chromosomes. The members of each pair are usually found together, as here.

The four pairs of chromosomes of...Drosophila are shown in Fig. 2. There are three pairs of large chromosomes and one pair of small chromosomes....

In Drosophila, then, there is a numerical correspondence between the number of hereditary groups and the number of the chromosomes. Moreover, the size relations of the groups and of the chromosomes correspond....

In most animals and plants the number of chromosomes is higher than in Drosophila, and the number of pairs of factors that may show independent assortment is, in consequence, increased. In the snail, Helix hortensis, the half number of the chromosomes is given as 22; in the potato beetle 18; in man, probably 24; in the mouse 20; in cotton 28;...in the garden pea 7; in the nightshade 36; in tobacco 24; in the tomato 12; in wheat 8. If 20 pairs of chromosomes are present there will be over one million possible kinds of germ cells in the F_1 hybrid. The number of combinations that two such sets of germ cells may produce through fertilization is enormously greater. From this point of view we can understand the absence of identical individuals in such mixed types as the human race. The chance of identity is still further decreased since in addition there may be very large numbers of factors within each chromosome.

PRESENT PROGRESS

Biology is now at a stage perhaps comparable with that of chemistry at the time of Lavoisier. The experimental method has become possible and is leading to progress in many directions.

It should be noted that the Dutch botanist De Vries has modified Darwin's conception of variations: he has shown that small fluctuating variations are of less importance in heredity than sudden and more noticeable mutations or "sports," perhaps due to the loss of a Mendelian factor. Mendel's work has been carried further with interesting results in the domains of stockbreeding and agriculture, notably the production of disease-resisting, heavy-cropping wheats. Loeb and other workers have shown what profound effects the environment may exert upon the organism, finding, for example, that the addition of magnesium salts to sea-water causes the two eyes of a developing fish to fuse into one. Much important work is also being done on the chemistry and physics of the organism itself.

It is now widely accepted that evolution occurred, that higher species arose by descent from lower species. We are endeavouring to understand the path which the evolutionary movement took, the forces which guided and controlled it. Our ultimate success may well be doubted. The expression of such doubt by the French philosopher Henri Bergson closes this collection of extracts.

HENRI BERGSON. CREATIVE EVOLUTION

(*English translation by Arthur Mitchell.*)

INTRODUCTION.

THE history of the evolution of life, incomplete as it yet is, already reveals to us how the intellect has been formed, by an uninterrupted progress, along a line which ascends through the vertebrate series up to man. It shows us in the faculty of understanding an appendage of the faculty of acting, a more and more precise, more and more complex and supple adaptation of the consciousness of living beings to the conditions of existence that are made for them. Hence should result this consequence that our intellect, in the narrow sense of the word, is intended to secure the perfect fitting of our body to its environment, to represent the relations of external things among themselves—in short, to think matter....We shall see that the human intellect feels at home among inanimate objects, more especially among solids, where our action finds its fulcrum and our industry its tools; that our concepts have been formed on the model of solids; that our logic is, pre-eminently, the logic of solids; that, consequently, our intellect triumphs in geometry, wherein is revealed the kinship of logical thought with unorganized matter, and where the intellect has only to follow its natural movement, after the lightest possible contact with experience, in order to go from discovery to discovery, sure that experience is following behind it and will justify it invariably.

But from this it must also follow that our thought, in its purely logical form, is incapable of presenting the true nature of life, the full meaning of the evolutionary movement. Created by life, in definite circumstances, to act on definite things, how can it embrace life, of which it is only an emanation or aspect? Deposited by the evolutionary movement in the course of its way, how can it be applied to the evolutionary movement itself? As well contend that the part is equal to the whole, that the effect can reabsorb its cause, or that the pebble left on the beach displays the form of the wave that brought it there. In fact, we do indeed feel that not one of the categories of our thought—unity, multiplicity, mechanical causality, intelligent finality, etc.

—applies exactly to the things of life: who can say where individuality begins and ends, whether the living being is one or many, whether it is the cells which associate themselves into the organism or the organism which dissociates itself into cells? In vain we force the living into this or that one of our moulds. All the moulds crack. They are too narrow, above all too rigid, for what we try to put into them. Our reasoning, so sure of itself among things inert, feels ill at ease on this new ground. It would be difficult to cite a biological discovery due to pure reasoning. And most often, when experience has finally shown us how life goes to work to obtain a certain result, we find its way of working is just that of which we should never have thought.

INDEX

Alexander the Great: taught by Aristotle, 3

Anaxagoras: held the geocentric theory, 33

Aquinas: welded Aristotelian philosophy and Christian dogma, 10

Arago: and the specific gravity of gases, 103

Archimedes: was in the line of immediate advance of Greek science, 3; wrote about Aristarchus, 7; on the Sand Reckoner, **8**

Aristarchus: was in the line of immediate advance of Greek science, 3; on the Sizes and Distances of the Sun and Moon, **7**; his work discussed by Archimedes, 9, not accepted by later astronomers, 10

Aristotle: an important Greek philosopher, 3; on the Heavens, **4**; the recovery of his work about 1220, 10; the welding of it with Christian dogma by Aquinas, 10, his biology, however, excepted, 188; his conception of an "unmoved mover," 14; his idea of a substance essentially light in nature, 83, 93; extract from Historia Animalium, **168**; from De Generatione Animalium, **170**; some of his facts used by Pliny, 171, 176; praised by Francis Bacon, 171; re-read at the Renaissance, 181

Arrhenius: on the dissociation of substances dissolved in water, **127**

Aston: and the Nobel Prize, 144; on Isotopes and Atomic Weights, **145**

Augustine: harmonized Greek evolutionary philosophy with Hebrew creation stories, 187

Avogadro: on molecules, **106**

Bacon (Francis): on natural history, 171

Badovere: wrote to Galileo about a telescope, 17

Baer, von: on the limbs of vertebrates, 242

Bateson: translation of Mendel's work, 247; discovery of linkage of Mendelian factors, 265

Becquerel: discovered radio-activity, 160

Bergson: extract from Creative Evolution, **268**

Berthollet: on the combination proportions of substances, 101 *et seq.*

Bieler: and α-particles, 164

Biot: and the specific gravity of gases, 103; friendship with Pasteur, 218

Bischoff: compared the structure of the brain in men and monkeys, 240, 242

Black: did experiments on combustion, 83

Bohr: his theory of atomic structure mentioned, 151

Borelli: a theory of impulse or attraction, 34

Boyle: used the atomic theory, 93; employed Hooke, 181

Bragg (W. H. and W. L.): examination of X-rays by means of crystals, 149; mentioned by Moseley, 150, 153

Brahe (Tycho): and Kepler, 31; his estimate of the distance of the moon, 38

Broek, van den: and the positive charge on atomic nuclei, 159, 160

Buckland: teacher of Lyell, 206

Bullialdus: on the distances of the planets from the sun, 37, 38; on the distance of the moon, 38

Bunsen: worked with Kirchhoff, 53 et seq.

A CATALOG OF SELECTED

DOVER BOOKS

IN ALL FIELDS OF INTEREST

STICKLEY CRAFTSMAN FURNITURE CATALOGS, Gustav Stickley and L. & J. G. Stickley. Beautiful, functional furniture in two authentic catalogs from 1910. 594 illustrations, including 277 photos, show settles, rockers, armchairs, reclining chairs, bookcases, desks, tables. 183pp. 6½ x 9¼. 23838-5

AMERICAN LOCOMOTIVES IN HISTORIC PHOTOGRAPHS: 1858 to 1949, Ron Ziel (ed.). A rare collection of 126 meticulously detailed official photographs, called "builder portraits," of American locomotives that majestically chronicle the rise of steam locomotive power in America. Introduction. Detailed captions. xi+ 129pp. 9 x 12. 27393-8

AMERICA'S LIGHTHOUSES: An Illustrated History, Francis Ross Holland, Jr. Delightfully written, profusely illustrated fact-filled survey of over 200 American lighthouses since 1716. History, anecdotes, technological advances, more. 240pp. 8 x 10¾. 25576-X

TOWARDS A NEW ARCHITECTURE, Le Corbusier. Pioneering manifesto by founder of "International School." Technical and aesthetic theories, views of industry, economics, relation of form to function, "mass-production split" and much more. Profusely illustrated. 320pp. 6⅛ x 9¼. (Available in U.S. only.) 25023-7

HOW THE OTHER HALF LIVES, Jacob Riis. Famous journalistic record, exposing poverty and degradation of New York slums around 1900, by major social reformer. 100 striking and influential photographs. 233pp. 10 x 7⅞. 22012-5

FRUIT KEY AND TWIG KEY TO TREES AND SHRUBS, William M. Harlow. One of the handiest and most widely used identification aids. Fruit key covers 120 deciduous and evergreen species; twig key 160 deciduous species. Easily used. Over 300 photographs. 126pp. 5⅜ x 8½. 20511-8

COMMON BIRD SONGS, Dr. Donald J. Borror. Songs of 60 most common U.S. birds: robins, sparrows, cardinals, bluejays, finches, more–arranged in order of increasing complexity. Up to 9 variations of songs of each species. Cassette and manual 99911-4

ORCHIDS AS HOUSE PLANTS, Rebecca Tyson Northen. Grow cattleyas and many other kinds of orchids–in a window, in a case, or under artificial light. 63 illustrations. 148pp. 5⅜ x 8½. 23261-1

MONSTER MAZES, Dave Phillips. Masterful mazes at four levels of difficulty. Avoid deadly perils and evil creatures to find magical treasures. Solutions for all 32 exciting illustrated puzzles. 48pp. 8¼ x 11. 26005-4

MOZART'S DON GIOVANNI (DOVER OPERA LIBRETTO SERIES), Wolfgang Amadeus Mozart. Introduced and translated by Ellen H. Bleiler. Standard Italian libretto, with complete English translation. Convenient and thoroughly portable–an ideal companion for reading along with a recording or the performance itself. Introduction. List of characters. Plot summary. 121pp. 5¼ x 8½. 24944-1

TECHNICAL MANUAL AND DICTIONARY OF CLASSICAL BALLET, Gail Grant. Defines, explains, comments on steps, movements, poses and concepts. 15-page pictorial section. Basic book for student, viewer. 127pp. 5⅜ x 8½. 21843-0

THE CLARINET AND CLARINET PLAYING, David Pino. Lively, comprehensive work features suggestions about technique, musicianship, and musical interpretation, as well as guidelines for teaching, making your own reeds, and preparing for public performance. Includes an intriguing look at clarinet history. "A godsend," *The Clarinet,* Journal of the International Clarinet Society. Appendixes. 7 illus. 320pp. $5^3/_8$ x $8^1/_2$. 40270-3

HOLLYWOOD GLAMOR PORTRAITS, John Kobal (ed.). 145 photos from 1926-49. Harlow, Gable, Bogart, Bacall; 94 stars in all. Full background on photographers, technical aspects. 160pp. $8^3/_8$ x $11^1/_4$. 23352-9

THE ANNOTATED CASEY AT THE BAT: A Collection of Ballads about the Mighty Casey/Third, Revised Edition, Martin Gardner (ed.). Amusing sequels and parodies of one of America's best-loved poems: Casey's Revenge, Why Casey Whiffed, Casey's Sister at the Bat, others. 256pp. $5^3/_8$ x $8^1/_2$. 28598-7

THE RAVEN AND OTHER FAVORITE POEMS, Edgar Allan Poe. Over 40 of the author's most memorable poems: "The Bells," "Ulalume," "Israfel," "To Helen," "The Conqueror Worm," "Eldorado," "Annabel Lee," many more. Alphabetic lists of titles and first lines. 64pp. $5^3/_{16}$ x $8^1/_4$. 26685-0

PERSONAL MEMOIRS OF U. S. GRANT, Ulysses Simpson Grant. Intelligent, deeply moving firsthand account of Civil War campaigns, considered by many the finest military memoirs ever written. Includes letters, historic photographs, maps and more. 528pp. $6^1/_8$ x $9^1/_4$. 28587-1

ANCIENT EGYPTIAN MATERIALS AND INDUSTRIES, A. Lucas and J. Harris. Fascinating, comprehensive, thoroughly documented text describes this ancient civilization's vast resources and the processes that incorporated them in daily life, including the use of animal products, building materials, cosmetics, perfumes and incense, fibers, glazed ware, glass and its manufacture, materials used in the mummification process, and much more. 544pp. $6^1/_8$ x $9^1/_4$. (Available in U.S. only.) 40446-3

RUSSIAN STORIES/RUSSKIE RASSKAZY: A Dual-Language Book, edited by Gleb Struve. Twelve tales by such masters as Chekhov, Tolstoy, Dostoevsky, Pushkin, others. Excellent word-for-word English translations on facing pages, plus teaching and study aids, Russian/English vocabulary, biographical/critical introductions, more. 416pp. $5^3/_8$ x $8^1/_2$. 26244-8

PHILADELPHIA THEN AND NOW: 60 Sites Photographed in the Past and Present, Kenneth Finkel and Susan Oyama. Rare photographs of City Hall, Logan Square, Independence Hall, Betsy Ross House, other landmarks juxtaposed with contemporary views. Captures changing face of historic city. Introduction. Captions. 128pp. $8^1/_4$ x 11. 25790-8

AIA ARCHITECTURAL GUIDE TO NASSAU AND SUFFOLK COUNTIES, LONG ISLAND, The American Institute of Architects, Long Island Chapter, and the Society for the Preservation of Long Island Antiquities. Comprehensive, well-researched and generously illustrated volume brings to life over three centuries of Long Island's great architectural heritage. More than 240 photographs with authoritative, extensively detailed captions. 176pp. $8^1/_4$ x 11. 26946-9

NORTH AMERICAN INDIAN LIFE: Customs and Traditions of 23 Tribes, Elsie Clews Parsons (ed.). 27 fictionalized essays by noted anthropologists examine religion, customs, government, additional facets of life among the Winnebago, Crow, Zuni, Eskimo, other tribes. 480pp. $6^1/_8$ x $9^1/_4$. 27377-6

THE BEST TALES OF HOFFMANN, E. T. A. Hoffmann. 10 of Hoffmann's most important stories: "Nutcracker and the King of Mice," "The Golden Flowerpot," etc. 458pp. 5⅜ x 8½. 21793-0

FROM FETISH TO GOD IN ANCIENT EGYPT, E. A. Wallis Budge. Rich detailed survey of Egyptian conception of "God" and gods, magic, cult of animals, Osiris, more. Also, superb English translations of hymns and legends. 240 illustrations. 545pp. 5⅜ x 8½. 25803-3

FRENCH STORIES/CONTES FRANÇAIS: A Dual-Language Book, Wallace Fowlie. Ten stories by French masters, Voltaire to Camus: "Micromegas" by Voltaire; "The Atheist's Mass" by Balzac; "Minuet" by de Maupassant; "The Guest" by Camus, six more. Excellent English translations on facing pages. Also French-English vocabulary list, exercises, more. 352pp. 5⅜ x 8½. 26443-2

CHICAGO AT THE TURN OF THE CENTURY IN PHOTOGRAPHS: 122 Historic Views from the Collections of the Chicago Historical Society, Larry A. Viskochil. Rare large-format prints offer detailed views of City Hall, State Street, the Loop, Hull House, Union Station, many other landmarks, circa 1904-1913. Introduction. Captions. Maps. 144pp. 9⅜ x 12¼. 24656-6

OLD BROOKLYN IN EARLY PHOTOGRAPHS, 1865-1929, William Lee Younger. Luna Park, Gravesend race track, construction of Grand Army Plaza, moving of Hotel Brighton, etc. 157 previously unpublished photographs. 165pp. 8⅞ x 11¾. 23587-4

THE MYTHS OF THE NORTH AMERICAN INDIANS, Lewis Spence. Rich anthology of the myths and legends of the Algonquins, Iroquois, Pawnees and Sioux, prefaced by an extensive historical and ethnological commentary. 36 illustrations. 480pp. 5⅜ x 8½. 25967-6

AN ENCYCLOPEDIA OF BATTLES: Accounts of Over 1,560 Battles from 1479 B.C. to the Present, David Eggenberger. Essential details of every major battle in recorded history from the first battle of Megiddo in 1479 B.C. to Grenada in 1984. List of Battle Maps. New Appendix covering the years 1967-1984. Index. 99 illustrations. 544pp. 6½ x 9¼. 24913-1

SAILING ALONE AROUND THE WORLD, Captain Joshua Slocum. First man to sail around the world, alone, in small boat. One of great feats of seamanship told in delightful manner. 67 illustrations. 294pp. 5⅜ x 8½. 20326-3

ANARCHISM AND OTHER ESSAYS, Emma Goldman. Powerful, penetrating, prophetic essays on direct action, role of minorities, prison reform, puritan hypocrisy, violence, etc. 271pp. 5⅜ x 8½. 22484-8

MYTHS OF THE HINDUS AND BUDDHISTS, Ananda K. Coomaraswamy and Sister Nivedita. Great stories of the epics; deeds of Krishna, Shiva, taken from puranas, Vedas, folk tales; etc. 32 illustrations. 400pp. 5⅜ x 8½. 21759-0

THE TRAUMA OF BIRTH, Otto Rank. Rank's controversial thesis that anxiety neurosis is caused by profound psychological trauma which occurs at birth. 256pp. 5⅜ x 8½. 27974-X

A THEOLOGICO-POLITICAL TREATISE, Benedict Spinoza. Also contains unfinished Political Treatise. Great classic on religious liberty, theory of government on common consent. R. Elwes translation. Total of 421pp. 5⅜ x 8½. 20249-6

PERSPECTIVE FOR ARTISTS, Rex Vicat Cole. Depth, perspective of sky and sea, shadows, much more, not usually covered. 391 diagrams, 81 reproductions of drawings and paintings. 279pp. 5⅜ x 8½. 22487-2

DRAWING THE LIVING FIGURE, Joseph Sheppard. Innovative approach to artistic anatomy focuses on specifics of surface anatomy, rather than muscles and bones. Over 170 drawings of live models in front, back and side views, and in widely varying poses. Accompanying diagrams. 177 illustrations. Introduction. Index. 144pp. 8⅜ x11¼. 26723-7

GOTHIC AND OLD ENGLISH ALPHABETS: 100 Complete Fonts, Dan X. Solo. Add power, elegance to posters, signs, other graphics with 100 stunning copyright-free alphabets: Blackstone, Dolbey, Germania, 97 more–including many lower-case, numerals, punctuation marks. 104pp. 8⅛ x 11. 24695-7

HOW TO DO BEADWORK, Mary White. Fundamental book on craft from simple projects to five-bead chains and woven works. 106 illustrations. 142pp. 5⅜ x 8. 20697-1

THE BOOK OF WOOD CARVING, Charles Marshall Sayers. Finest book for beginners discusses fundamentals and offers 34 designs. "Absolutely first rate . . . well thought out and well executed."–E. J. Tangerman. 118pp. 7¾ x 10⅝. 23654-4

ILLUSTRATED CATALOG OF CIVIL WAR MILITARY GOODS: Union Army Weapons, Insignia, Uniform Accessories, and Other Equipment, Schuyler, Hartley, and Graham. Rare, profusely illustrated 1846 catalog includes Union Army uniform and dress regulations, arms and ammunition, coats, insignia, flags, swords, rifles, etc. 226 illustrations. 160pp. 9 x 12. 24939-5

WOMEN'S FASHIONS OF THE EARLY 1900s: An Unabridged Republication of "New York Fashions, 1909," National Cloak & Suit Co. Rare catalog of mail-order fashions documents women's and children's clothing styles shortly after the turn of the century. Captions offer full descriptions, prices. Invaluable resource for fashion, costume historians. Approximately 725 illustrations. 128pp. 8⅜ x 11¼. 27276-1

THE 1912 AND 1915 GUSTAV STICKLEY FURNITURE CATALOGS, Gustav Stickley. With over 200 detailed illustrations and descriptions, these two catalogs are essential reading and reference materials and identification guides for Stickley furniture. Captions cite materials, dimensions and prices. 112pp. 6½ x 9¼. 26676-1

EARLY AMERICAN LOCOMOTIVES, John H. White, Jr. Finest locomotive engravings from early 19th century: historical (1804–74), main-line (after 1870), special, foreign, etc. 147 plates. 142pp. 11⅜ x 8¼. 22772-3

THE TALL SHIPS OF TODAY IN PHOTOGRAPHS, Frank O. Braynard. Lavishly illustrated tribute to nearly 100 majestic contemporary sailing vessels: Amerigo Vespucci, Clearwater, Constitution, Eagle, Mayflower, Sea Cloud, Victory, many more. Authoritative captions provide statistics, background on each ship. 190 black-and-white photographs and illustrations. Introduction. 128pp. 8⅞ x 11¾. 27163-3

LITTLE BOOK OF EARLY AMERICAN CRAFTS AND TRADES, Peter Stockham (ed.). 1807 children's book explains crafts and trades: baker, hatter, cooper, potter, and many others. 23 copperplate illustrations. 140pp. 4⅝ x 6. 23336-7

VICTORIAN FASHIONS AND COSTUMES FROM HARPER'S BAZAR, 1867–1898, Stella Blum (ed.). Day costumes, evening wear, sports clothes, shoes, hats, other accessories in over 1,000 detailed engravings. 320pp. 9⅜ x 12¼. 22990-4

GUSTAV STICKLEY, THE CRAFTSMAN, Mary Ann Smith. Superb study surveys broad scope of Stickley's achievement, especially in architecture. Design philosophy, rise and fall of the Craftsman empire, descriptions and floor plans for many Craftsman houses, more. 86 black-and-white halftones. 31 line illustrations. Introduction 208pp. 6½ x 9¼. 27210-9

THE LONG ISLAND RAIL ROAD IN EARLY PHOTOGRAPHS, Ron Ziel. Over 220 rare photos, informative text document origin (1844) and development of rail service on Long Island. Vintage views of early trains, locomotives, stations, passengers, crews, much more. Captions. 8⅞ x 11¾. 26301-0

VOYAGE OF THE LIBERDADE, Joshua Slocum. Great 19th-century mariner's thrilling, first-hand account of the wreck of his ship off South America, the 35-foot boat he built from the wreckage, and its remarkable voyage home. 128pp. 5⅜ x 8½. 40022-0

TEN BOOKS ON ARCHITECTURE, Vitruvius. The most important book ever written on architecture. Early Roman aesthetics, technology, classical orders, site selection, all other aspects. Morgan translation. 331pp. 5⅜ x 8½. 20645-9

THE HUMAN FIGURE IN MOTION, Eadweard Muybridge. More than 4,500 stopped-action photos, in action series, showing undraped men, women, children jumping, lying down, throwing, sitting, wrestling, carrying, etc. 390pp. 7⅞ x 10⅝. 20204-6 Clothbd.

TREES OF THE EASTERN AND CENTRAL UNITED STATES AND CANADA, William M. Harlow. Best one-volume guide to 140 trees. Full descriptions, woodlore, range, etc. Over 600 illustrations. Handy size. 288pp. 4½ x 6⅜. 20395-6

SONGS OF WESTERN BIRDS, Dr. Donald J. Borror. Complete song and call repertoire of 60 western species, including flycatchers, juncoes, cactus wrens, many more–includes fully illustrated booklet. Cassette and manual 99913-0

GROWING AND USING HERBS AND SPICES, Milo Miloradovich. Versatile handbook provides all the information needed for cultivation and use of all the herbs and spices available in North America. 4 illustrations. Index. Glossary. 236pp. 5⅜ x 8½. 25058-X

BIG BOOK OF MAZES AND LABYRINTHS, Walter Shepherd. 50 mazes and labyrinths in all–classical, solid, ripple, and more–in one great volume. Perfect inexpensive puzzler for clever youngsters. Full solutions. 112pp. 8⅛ x 11. 22951-3

PIANO TUNING, J. Cree Fischer. Clearest, best book for beginner, amateur. Simple repairs, raising dropped notes, tuning by easy method of flattened fifths. No previous skills needed. 4 illustrations. 201pp. 5⅜ x 8½. 23267-0

HINTS TO SINGERS, Lillian Nordica. Selecting the right teacher, developing confidence, overcoming stage fright, and many other important skills receive thoughtful discussion in this indispensible guide, written by a world-famous diva of four decades' experience. 96pp. 5⅜ x 8½. 40094-8

THE COMPLETE NONSENSE OF EDWARD LEAR, Edward Lear. All nonsense limericks, zany alphabets, Owl and Pussycat, songs, nonsense botany, etc., illustrated by Lear. Total of 320pp. 5⅜ x 8½. (Available in U.S. only.) 20167-8

VICTORIAN PARLOUR POETRY: An Annotated Anthology, Michael R. Turner. 117 gems by Longfellow, Tennyson, Browning, many lesser-known poets. "The Village Blacksmith," "Curfew Must Not Ring Tonight," "Only a Baby Small," dozens more, often difficult to find elsewhere. Index of poets, titles, first lines. xxiii + 325pp. 5⅝ x 8¼. 27044-0

DUBLINERS, James Joyce. Fifteen stories offer vivid, tightly focused observations of the lives of Dublin's poorer classes. At least one, "The Dead," is considered a masterpiece. Reprinted complete and unabridged from standard edition. 160pp. 5$\frac{3}{16}$ x 8¼. 26870-5

GREAT WEIRD TALES: 14 Stories by Lovecraft, Blackwood, Machen and Others, S. T. Joshi (ed.). 14 spellbinding tales, including "The Sin Eater," by Fiona McLeod, "The Eye Above the Mantel," by Frank Belknap Long, as well as renowned works by R. H. Barlow, Lord Dunsany, Arthur Machen, W. C. Morrow and eight other masters of the genre. 256pp. 5⅜ x 8½. (Available in U.S. only.) 40436-6

THE BOOK OF THE SACRED MAGIC OF ABRAMELIN THE MAGE, translated by S. MacGregor Mathers. Medieval manuscript of ceremonial magic. Basic document in Aleister Crowley, Golden Dawn groups. 268pp. 5⅜ x 8½. 23211-5

NEW RUSSIAN-ENGLISH AND ENGLISH-RUSSIAN DICTIONARY, M. A. O'Brien. This is a remarkably handy Russian dictionary, containing a surprising amount of information, including over 70,000 entries. 366pp. 4½ x 6⅛. 20208-9

HISTORIC HOMES OF THE AMERICAN PRESIDENTS, Second, Revised Edition, Irvin Haas. A traveler's guide to American Presidential homes, most open to the public, depicting and describing homes occupied by every American President from George Washington to George Bush. With visiting hours, admission charges, travel routes. 175 photographs. Index. 160pp. 8¼ x 11. 26751-2

NEW YORK IN THE FORTIES, Andreas Feininger. 162 brilliant photographs by the well-known photographer, formerly with *Life* magazine. Commuters, shoppers, Times Square at night, much else from city at its peak. Captions by John von Hartz. 181pp. 9¼ x 10¾. 23585-8

INDIAN SIGN LANGUAGE, William Tomkins. Over 525 signs developed by Sioux and other tribes. Written instructions and diagrams. Also 290 pictographs. 111pp. 6⅛ x 9¼. 22029-X

ANATOMY: A Complete Guide for Artists, Joseph Sheppard. A master of figure drawing shows artists how to render human anatomy convincingly. Over 460 illustrations. 224pp. 8⅜ x 11¼. 27279-6

MEDIEVAL CALLIGRAPHY: Its History and Technique, Marc Drogin. Spirited history, comprehensive instruction manual covers 13 styles (ca. 4th century through 15th). Excellent photographs; directions for duplicating medieval techniques with modern tools. 224pp. 8⅜ x 11¼. 26142-5

DRIED FLOWERS: How to Prepare Them, Sarah Whitlock and Martha Rankin. Complete instructions on how to use silica gel, meal and borax, perlite aggregate, sand and borax, glycerine and water to create attractive permanent flower arrangements. 12 illustrations. 32pp. 5⅜ x 8½. 21802-3

EASY-TO-MAKE BIRD FEEDERS FOR WOODWORKERS, Scott D. Campbell. Detailed, simple-to-use guide for designing, constructing, caring for and using feeders. Text, illustrations for 12 classic and contemporary designs. 96pp. 5⅜ x 8½. 25847-5

SCOTTISH WONDER TALES FROM MYTH AND LEGEND, Donald A. Mackenzie. 16 lively tales tell of giants rumbling down mountainsides, of a magic wand that turns stone pillars into warriors, of gods and goddesses, evil hags, powerful forces and more. 240pp. 5⅜ x 8½. 29677-6

THE HISTORY OF UNDERCLOTHES, C. Willett Cunnington and Phyllis Cunnington. Fascinating, well-documented survey covering six centuries of English undergarments, enhanced with over 100 illustrations: 12th-century laced-up bodice, footed long drawers (1795), 19th-century bustles, 19th-century corsets for men, Victorian "bust improvers," much more. 272pp. 5⅜ x 8¼. 27124-2

ARTS AND CRAFTS FURNITURE: The Complete Brooks Catalog of 1912, Brooks Manufacturing Co. Photos and detailed descriptions of more than 150 now very collectible furniture designs from the Arts and Crafts movement depict davenports, settees, buffets, desks, tables, chairs, bedsteads, dressers and more, all built of solid, quarter-sawed oak. Invaluable for students and enthusiasts of antiques, Americana and the decorative arts. 80pp. 6½ x 9¼. 27471-3

WILBUR AND ORVILLE: A Biography of the Wright Brothers, Fred Howard. Definitive, crisply written study tells the full story of the brothers' lives and work. A vividly written biography, unparalleled in scope and color, that also captures the spirit of an extraordinary era. 560pp. 6⅛ x 9¼. 40297-5

THE ARTS OF THE SAILOR: Knotting, Splicing and Ropework, Hervey Garrett Smith. Indispensable shipboard reference covers tools, basic knots and useful hitches; handsewing and canvas work, more. Over 100 illustrations. Delightful reading for sea lovers. 256pp. 5⅜ x 8½. 26440-8

FRANK LLOYD WRIGHT'S FALLINGWATER: The House and Its History, Second, Revised Edition, Donald Hoffmann. A total revision–both in text and illustrations–of the standard document on Fallingwater, the boldest, most personal architectural statement of Wright's mature years, updated with valuable new material from the recently opened Frank Lloyd Wright Archives. "Fascinating"–*The New York Times*. 116 illustrations. 128pp. 9¼ x 10¾. 27430-6

THE WIT AND HUMOR OF OSCAR WILDE, Alvin Redman (ed.). More than 1,000 ripostes, paradoxes, wisecracks: Work is the curse of the drinking classes; I can resist everything except temptation; etc. 258pp. 5⅜ x 8½. 20602-5

SHAKESPEARE LEXICON AND QUOTATION DICTIONARY, Alexander Schmidt. Full definitions, locations, shades of meaning in every word in plays and poems. More than 50,000 exact quotations. 1,485pp. 6½ x 9¼. 2-vol. set.
Vol. 1: 22726-X
Vol. 2: 22727-8

SELECTED POEMS, Emily Dickinson. Over 100 best-known, best-loved poems by one of America's foremost poets, reprinted from authoritative early editions. No comparable edition at this price. Index of first lines. 64pp. 5$\frac{3}{16}$ x 8¼. 26466-1

THE INSIDIOUS DR. FU-MANCHU, Sax Rohmer. The first of the popular mystery series introduces a pair of English detectives to their archnemesis, the diabolical Dr. Fu-Manchu. Flavorful atmosphere, fast-paced action, and colorful characters enliven this classic of the genre. 208pp. 5$\frac{3}{16}$ x 8¼. 29898-1

THE MALLEUS MALEFICARUM OF KRAMER AND SPRENGER, translated by Montague Summers. Full text of most important witchhunter's "bible," used by both Catholics and Protestants. 278pp. 6⅝ x 10. 22802-9

SPANISH STORIES/CUENTOS ESPAÑOLES: A Dual-Language Book, Angel Flores (ed.). Unique format offers 13 great stories in Spanish by Cervantes, Borges, others. Faithful English translations on facing pages. 352pp. 5⅜ x 8½. 25399-6

GARDEN CITY, LONG ISLAND, IN EARLY PHOTOGRAPHS, 1869–1919, Mildred H. Smith. Handsome treasury of 118 vintage pictures, accompanied by carefully researched captions, document the Garden City Hotel fire (1899), the Vanderbilt Cup Race (1908), the first airmail flight departing from the Nassau Boulevard Aerodrome (1911), and much more. 96pp. 8⅞ x 11¾. 40669-5

OLD QUEENS, N.Y., IN EARLY PHOTOGRAPHS, Vincent F. Seyfried and William Asadorian. Over 160 rare photographs of Maspeth, Jamaica, Jackson Heights, and other areas. Vintage views of DeWitt Clinton mansion, 1939 World's Fair and more. Captions. 192pp. 8⅞ x 11. 26358-4

CAPTURED BY THE INDIANS: 15 Firsthand Accounts, 1750-1870, Frederick Drimmer. Astounding true historical accounts of grisly torture, bloody conflicts, relentless pursuits, miraculous escapes and more, by people who lived to tell the tale. 384pp. 5⅜ x 8½. 24901-8

THE WORLD'S GREAT SPEECHES (Fourth Enlarged Edition), Lewis Copeland, Lawrence W. Lamm, and Stephen J. McKenna. Nearly 300 speeches provide public speakers with a wealth of updated quotes and inspiration–from Pericles' funeral oration and William Jennings Bryan's "Cross of Gold Speech" to Malcolm X's powerful words on the Black Revolution and Earl of Spenser's tribute to his sister, Diana, Princess of Wales. 944pp. 5⅜ x 8⅜. 40903-1

THE BOOK OF THE SWORD, Sir Richard F. Burton. Great Victorian scholar/adventurer's eloquent, erudite history of the "queen of weapons"–from prehistory to early Roman Empire. Evolution and development of early swords, variations (sabre, broadsword, cutlass, scimitar, etc.), much more. 336pp. 6⅛ x 9¼.
25434-8

AUTOBIOGRAPHY: The Story of My Experiments with Truth, Mohandas K. Gandhi. Boyhood, legal studies, purification, the growth of the Satyagraha (nonviolent protest) movement. Critical, inspiring work of the man responsible for the freedom of India. 480pp. 5⅜ x 8½. (Available in U.S. only.) 24593-4

CELTIC MYTHS AND LEGENDS, T. W. Rolleston. Masterful retelling of Irish and Welsh stories and tales. Cuchulain, King Arthur, Deirdre, the Grail, many more. First paperback edition. 58 full-page illustrations. 512pp. 5⅜ x 8½. 26507-2

THE PRINCIPLES OF PSYCHOLOGY, William James. Famous long course complete, unabridged. Stream of thought, time perception, memory, experimental methods; great work decades ahead of its time. 94 figures. 1,391pp. 5⅜ x 8½. 2-vol. set.
Vol. I: 20381-6 Vol. II: 20382-4

THE WORLD AS WILL AND REPRESENTATION, Arthur Schopenhauer. Definitive English translation of Schopenhauer's life work, correcting more than 1,000 errors, omissions in earlier translations. Translated by E. F. J. Payne. Total of 1,269pp. 5⅜ x 8½. 2-vol. set. Vol. 1: 21761-2 Vol. 2: 21762-0

MAGIC AND MYSTERY IN TIBET, Madame Alexandra David-Neel. Experiences among lamas, magicians, sages, sorcerers, Bonpa wizards. A true psychic discovery. 32 illustrations. 321pp. 5⅜ x 8½. (Available in U.S. only.) 22682-4

THE EGYPTIAN BOOK OF THE DEAD, E. A. Wallis Budge. Complete reproduction of Ani's papyrus, finest ever found. Full hieroglyphic text, interlinear transliteration, word-for-word translation, smooth translation. 533pp. 6½ x 9¼. 21866-X

MATHEMATICS FOR THE NONMATHEMATICIAN, Morris Kline. Detailed, college-level treatment of mathematics in cultural and historical context, with numerous exercises. Recommended Reading Lists. Tables. Numerous figures. 641pp. 5⅜ x 8½.
24823-2

PROBABILISTIC METHODS IN THE THEORY OF STRUCTURES, Isaac Elishakoff. Well-written introduction covers the elements of the theory of probability from two or more random variables, the reliability of such multivariable structures, the theory of random function, Monte Carlo methods of treating problems incapable of exact solution, and more. Examples. 502pp. 5⅜ x 8½. 40691-1

THE RIME OF THE ANCIENT MARINER, Gustave Doré, S. T. Coleridge. Doré's finest work; 34 plates capture moods, subtleties of poem. Flawless full-size reproductions printed on facing pages with authoritative text of poem. "Beautiful. Simply beautiful."–*Publisher's Weekly*. 77pp. 9¼ x 12. 22305-1

NORTH AMERICAN INDIAN DESIGNS FOR ARTISTS AND CRAFTSPEOPLE, Eva Wilson. Over 360 authentic copyright-free designs adapted from Navajo blankets, Hopi pottery, Sioux buffalo hides, more. Geometrics, symbolic figures, plant and animal motifs, etc. 128pp. 8⅜ x 11. (Not for sale in the United Kingdom.) 25341-4

SCULPTURE: Principles and Practice, Louis Slobodkin. Step-by-step approach to clay, plaster, metals, stone; classical and modern. 253 drawings, photos. 255pp. 8⅛ x 11.
22960-2

THE INFLUENCE OF SEA POWER UPON HISTORY, 1660–1783, A. T. Mahan. Influential classic of naval history and tactics still used as text in war colleges. First paperback edition. 4 maps. 24 battle plans. 640pp. 5⅜ x 8½. 25509-3